*Encyclopaedia of*

# World Geography

*Encyclopaedia of*
# World Geography

## Vol. - 1

Chandra Vijay Purty

**ABD PUBLISHERS**
Jaipur ■ New Delhi

**ABD PUBLISHERS**

**Regd. Off.** "Bony Residency", Gate No. 2
Opp. Tilak Public School, Vishweswariya Nagar,
Gopalpura Road, Jaipur - 302018, India
Ph: 0141 - 2761280, Telefax: 2761381
Email: abdpublishersjpr@gmail.com

**Delhi Office:** 102, 1st Floor, Satyam House,
4327/3, Ansari Road, Darya Ganj,
New Delhi - 110002, India
Ph: 011-45652440

visit us: www.oxfordbookcompany.com

Published in India, 2024



ISBN : 9788183763127 (7 Vols. Set)



Typeset by Shivangi Computer, Jaipur

Printed and bound in India by Thomson Press Ltd., New Delhi

# Preface

Geography is the study of the Earth and its lands, features, inhabitants, and phenomena. A literal translation would be "to describe or write about the Earth". The first person to use the word "geography" was Eratosthenes (276-194 B.C.). Four historical traditions in geographical research are the spatial analysis of natural and human phenomena (geography as a study of distribution), area studies (places and regions), study of man-land relationship, and research in earth sciences. Nonetheless, modern geography is an all-encompassing discipline that foremost seeks to understand the Earth and all of its human and natural complexities—not merely where objects are, but how they have changed and come to be. As "the bridge between the human and physical sciences," geography is divided into two main branches—human geography and physical geography.

The oldest known world maps date back to ancient Babylon from the 9th century BC. The best known Babylonian world map, however, is the *Imago Mundi* of 600 BC. The map as reconstructed by Eckhard Unger shows Babylon on the Euphrates, surrounded by a circular landmass showing Assyria, Urartu and several cities, in turn surrounded by a "bitter river" (Oceanus), with seven islands arranged around it so as to form a seven-pointed star. The accompanying text mentions seven outer regions beyond the encircling ocean. The descriptions of five of them have survived. In contrast to the *Imago Mundi*, an earlier Babylonian world map dating back to the 9th century BC depicted Babylon as being further north from the centre of the world, though it is not certain what that centre was supposed to represent.

The ideas of Anaximander (c. 610 B.C.-c. 545 B.C.), considered by later Greek writers to be the true founder of geography, come to us through fragments quoted by his successors. Anaximander is credited with the invention of the gnomon, the simple yet efficient Greek instrument that allowed the early measurement of latitude. Thales, Anaximander is also credited with the prediction of eclipses. The foundations of geography can be traced to the ancient cultures, such as the ancient, medieval, and early modern Chinese. The Greeks, who were the first to explore geography as both art and science, achieved this through Cartography, Philosophy, and Literature, or through Mathematics. There is some debate about who was the first person to assert that the Earth is spherical in shape, with the credit going either to Parmenides or Pythagoras. Anaxagoras was able to demonstrate that the profile of the Earth was circular by explaining eclipses. However, he still believed that the Earth was a flat disk, as did many of his contemporaries. One of the first estimates of the radius of the Earth was made by Eratosthenes.

The first rigorous system of latitude and longitude lines is credited to Hipparchus. He employed a sexagesimal system that was derived from Babylonian mathematics. The parallels and meridians were sub-divided into 360°, with each degree further subdivided 602 (minutes). To measure the longitude at different location on Earth, he suggested using eclipses to determine the relative difference in time. The extensive mapping by the Romans as they explored new lands would later provide a high level of information for Ptolemy to construct detailed atlases. He extended the work of Hipparchus, using a grid system on his maps and adopting a length of 56.5 miles for a degree.

From the 3rd century onwards, Chinese methods of geographical study and writing of geographical literature became much more complex than what was found in Europe at the time (until the 13th century). Chinese geographers such as Liu An, Pei Xiu, Jia Dan, Shen Kuo, Fan Chengda, Zhou Daguan, and Xu Xiake wrote important treatises, yet by the 17th century, advanced ideas and methods of Western-style geography were adopted in China.

This book contains the fundamental and basic information of the subject.

**Chandra Vijay Purty**

# Contents

# 1

# Introduction

Asia is the world's largest and most populous continent, located primarily in the eastern and northern hemispheres. It covers 8.6% of the Earth's total surface area (or 29.9% of its land area) and with approximately 4 billion people, it hosts 60% of the world's current human population. During the 20th century Asia's population nearly quadrupled.

Asia is traditionally defined as part of the landmass of Eurasia — with the western portion of the latter occupied by Europe — located to the east of the Suez Canal, east of the Ural Mountains and south of the Caucasus Mountains (or the Kuma-Manych Depression) and the Caspian and Black Seas. It is bounded on the east by the Pacific Ocean, on the south by the Indian Ocean and on the north by the Arctic Ocean. Given its size and diversity, Asia — a toponym dating back to classical antiquity — is more a cultural concept incorporating a number of regions and peoples than a homogeneous physical entity.

The wealth of Asia differs very widely among and within its regions, due to its vast size and huge range of different cultures, environments, historical ties and government systems. In terms of nominal GDP, Japan has the largest economy on the continent and the second largest in the world. In purchasing power parity terms, however, China has the largest economy in Asia and the second largest in the world.

## Etymology

The term "Asia" is originally a concept exclusively of Western civilization. The peoples of ancient *Asia* (Chinese, Japanese,

Indians, Persians, Arabs etc.) never conceived the idea of *Asia*, simply because they did not see themselves collectively. In their perspective, they were vastly varied civilizations, contrary to ancient European belief.

The word *Asia* originated from the Greek word *Ασιά*, first attributed to Herodotus (about 440 BC) in reference to Anatolia or — in describing the Persian Wars — to the Persian Empire, in contrast to Greece and Egypt. Herodotus comments that he is puzzled as to why three women's names are used to describe one enormous and substantial land mass (Europa, Asia, and Libya, referring to Africa), stating that most Greeks assumed that Asia was named after the wife of Prometheus (i.e. Hesione), but that the Lydians say it was named after Asias, son of Cotys, who passed the name on to a tribe in Sardis. Even before Herodotus.

Usage of the term soon became common in ancient Greece, and subsequently by the ancient Romans. Ancient and medieval European maps depict the Asian continent as a "huge amorphous blob" extending eastward. It was presumed in antiquity to end with India — the Greek king Alexander the Great believing he would reach the "end of the world" upon his arrival in the East.

### *Other Alternatives*

Alternatively, the etymology of the term may be from the Akkadian word *(w) aSû(m)*, which means 'to go outside' or 'to ascend', referring to the direction of the sun at sunrise in the Middle East and also likely connected with the Phoenician word *asa* meaning east. This may be contrasted to a similar etymology proposed for *Europe*, as being from Akkadian *erebu(m)* 'to enter' or 'set' (of the sun).

T.R. Reid supports this alternative etymology, noting that the ancient Greek name must have derived from *asu*, meaning 'east' in Assyrian (*ereb* for *Europe* meaning 'west'). The ideas of *Occidental* (form Latin *Occidens* 'setting') and *Oriental* (from Latin *Oriens* for 'rising') are also European invention, synonymous with *Western* and *Eastern*. Reid further emphasizes that it explains the Western point of view of placing all the peoples and cultures of Asia into a single classification, almost as if there were a need for setting the distinction between Western and Eastern civilizations on the Eurasian continent.

Ogura Kazuo and Tenshin Okakura are two Japanese outspoken figures over the subject.

However, this etymology is considered doubtful, because it does not explain how the term "Asia" first came to be associated with Anatolia, which is *west* of the Semitic-speaking areas, unless they refer to the viewpoint of a Phoenician sailor sailing through the straits between the Mediterranean Sea and the Black Sea.

## Definition and Boundaries

### *Physical Geography*

Medieval Europeans considered Asia as a continent a distinct landmass. The European concept of the three continents in the Old World goes back to Classical Antiquity, but during the Middle Ages was notably due to 7th century Spanish scholar Isidore of Sevilla. The demarcation between Asia and Africa (to the southwest) is the Isthmus of Suez and the Red Sea. The boundary between Asia and Europe is conventionally considered to run through the Dardanelles, the Sea of Marmara, the Bosporus, the Black Sea, the Caucasus Mountains, the Caspian Sea, the Ural River to its source and the Ural Mountains to the Kara Sea near Kara, Russia. While this interpretation of tripartite continents (i.e., of Asia, Europe and Africa) remains common in modernity, discovery of the extent of Africa and Asia have made this definition somewhat anachronistic. This is especially true in the case of Asia, which has several regions that would be considered distinct landmasses if these criteria were used (for example, Southern Asia and Eastern Asia).

In the far northeast of Asia, Siberia is separated from North America by the Bering Strait. Asia is bounded on the south by the Indian Ocean (specifically, from west to east, the Gulf of Aden, Arabian Sea and Bay of Bengal), on the east by the waters of the Pacific Ocean (including, counter clockwise, the South China Sea, East China Sea, Yellow Sea, Sea of Japan, Sea of Okhotsk and Bering Sea) and on the north by the Arctic Ocean. Australia (or Oceania) is to the southeast.

Some geographers do not consider Asia and Europe to be separate continents, as there is no logical physical separation between them. For example, Sir Barry Cunliffe, the emeritus professor of European archeology at Oxford, argues that Europe has been geographically and culturally merely "the western excrescence of the continent of Asia." Geographically, Asia is the major eastern constituent of the continent of Eurasia with Europe being a northwestern peninsula of the landmass – or of Afro-Eurasia: geologically, Asia, Europe and Africa comprise a single continuous landmass (save the Suez Canal) and share a common continental shelf. Almost all of Europe and most of Asia sit atop the Eurasian Plate, adjoined on the south by the Arabian and Indian Plate and with the easternmost part of Siberia (east of the Cherskiy Range) on the North American Plate.

In geography, there are two schools of thought. One school follows historical convention and treats Europe and Asia as different continents, categorizing subregions within them for more detailed analysis. The other school equates the word "continent" with a geographical region when referring to Europe, and use the term "region" to describe Asia in terms of physiography. Since, in linguistic terms, "continent" implies a distinct landmass, it is becoming increasingly common to substitute the term "region" for "continent" to avoid the problem of disambiguation altogether.

Given the scope and diversity of the landmass, it is sometimes not even clear exactly what "Asia" consists of. Some definitions exclude Turkey, the Middle East, Central Asia and Russia while only considering the Far East, Southeast Asia and the Indian subcontinent to compose Asia, especially in the United States after World War II. The term is sometimes used more narrowly in reference to the Asia-Pacific region, which does not include the Middle East, South Asia or Russia, but does include islands in the Pacific Ocean—a number of which may also be considered part of Australasia or Oceania, although Pacific Islanders are not considered Asian.

## Economy

The economy of Asia comprises more than 4 billion people (60% of the world population) living in 46 different states. Six further states lie partly in Asia, but are considered to belong to another region economically and politically. As in all world regions, the wealth of Asia differs widely between, and within, states. This is due to its vast size, meaning a huge range of differing cultures, environments, historical ties and government systems. The largest economies in Asia in terms of nominal GDP are Japan, China, India, South Korea, Indonesia and Iran. Economies range from Japan, as the world's second largest economy by nominal GDP, to Cambodia as one of the poorest. In terms of GDP by purchasing power parity, China has the largest economy in Asia and the second largest economy in the world, followed by Japan and India as the world's third and fourth largest economies respectively.

Wealth (if measured by GDP per capita) is mostly concentrated in east Asian territories such as Hong Kong, Japan, South Korea, and Taiwan, as well as in oil rich Middle Eastern countries such as Iran, Saudi Arabia, Qatar, United Arab Emirates. Asia, with the exception of Japan and South Korea, is currently undergoing rapid growth and industrialization spearheaded by China and India-the two fastest growing major economies in the world. While east Asian and southeast Asian countries generally rely on manufacturing and trade for growth, countries in the Middle East depend more on the production of commodities, principally oil, for economic growth. Over the years, with rapid economic growth and large trade surplus with the rest of the world, Asia has accumulated over US$4 trillion of foreign exchange reserves-more than half of the world's total.

## Economic Development

### *Ancient and Medieval Times*

Asia was rich in the ancient times. China and India alternated in being the largest economies in the world from about 1 A.D to 1800 A.D. China was a major economic power and attracted many to the east, and for many, the legendary wealth and prosperity of the ancient culture of India personified Asia, attracting European commerce, exploration and colonialism. The accidental discovery of America by Columbus in search for India demonstrates this deep fascination. The Silk Road became the main East-West trading route in the Asian hitherland while the Straits of Malacca stood as a major sea route.

***Pre-1945***

Prior to World War II, most of Asia was under colonial rule. Only relatively few states managed to remain independent in the face of constant pressure exerted by European power. Such examples are Siam and Japan.

Japan in particular managed to develop its economy due to a reformation in the 19th century. The reformation was comprehensive and is today known as the Meiji Restoration. The Japanese economy continued to grow well into the 20th century and its economic growth created various shortages of resources essential to economic growth. As a result the Japanese expansion began with a great part of Korea and China annexed, thus allowing the Japanese to secure strategic resources.

At the same time, Southeast Asia was prospering due to trade and the introduction of various new technologies of that time. The volume of trade continued to increase with the opening of the Suez Canal in the 1860s. Manila had its gallion or Manila galleon wherein products from the Philippines were traded to Europe. The Philippines was the first Asian country to trade with Latin America via Acapulco. Tobacco, coconut, corn, and sugar trade was the most in demand during that time. Singapore, founded in 1819, rose to prominence as trade between the east and the west increased at an incredible rate. The British colony of Malaya, now part of Malaysia, was the world's largest producer of tin and rubber. The Dutch East Indies, now Indonesia, on the other hand, was known for its spices production. Both the British and the Dutch created their own trading companies to manage their trade flow in Asia. The British created the British East India Company while the Dutch formed Dutch East India Company. Both companies maintained trade monopolies of their respective colonies.

In 1908, crude oil was first discovered in Persia, modern day Iran. Afterwards, many oil fields were discovered and it was learnt later that the Mideast possesses the world's largest oil stocks. This made the rulers of the Arab nations very rich though the socioeconomic development in that region lagged behind.

In the early 1930s, the world underwent a global economic depression, today known as the Great Depression. Asia was not spared, and suffered the same pain as Europe and the United States. The volume of trade decreased dramatically all around Asia and indeed the world. With falling demand, prices of various goods starting to fall and further impoverished locals and foreigners alike. In 1941, Japan invaded Malaya and thus began World War II in Asia.

### *1945-1990*

Following World War II, the People's Republic of China and India, which account for half of the population of Asia, adopted socialist policies to promote their domestic economy. These policies limited the economic growth of the region. In contrast, the economies of superiors Japan, South Korea and the other tigers Taiwan, Singapore, and Hong Kong—were economic successes, and the only successful economies outside of North America, Western Europe and Australia. The Philippines from the post-World War II until the late 1960s had the second largest economy in Asia. The Philippine economy during the 1980s was marked by stagnant growth as a result of dictatorship and martial law.

One of the most pronounced Asian economic phenomenons during this time-the Japanese post-war economic miracle greatly impacted the rest of the world. After World War II, under central guidance from the Japanese government, the entire economy was undergoing a remarkable restructuring. Close cooperation between the government, corporations and banks facilitated easy access to much-needed capital, and large conglomerates known as *keiretsu* spurred horizontal and vertical integration across all industries, keeping out foreign competition. These policies, in addition to an abandonment of military spending, worked phenomenally well. Japanese corporations as a result exported and still export massive amounts of high quality products from The Land of The Rising Sun.

Another amazing economic success story is that of South Korea's, also referred to as the Miracle on the Han River. The country was left impoverished after the Korean War, yet was able to recover at double digit percentiles. Many conglomerates, also known as Chaebols, such as Samsung, LG, Hyundai, Kia, SK, and more grew tremendously during this period. South Korea has now become the most wired country in the world.

Taiwan and Hong Kong experienced rapid growth up till the 1990s. Taiwan became, and still remains one of the main centers of consumer electronics R&D as well as manufacturing. However, unlike in Japan and South Korea, the bulk of Taiwan's economy is dependent on small to medium sized businesses. Hong Kong, on the other hand, experienced rapid growth in the financial sector due to liberal market policies, with many financial institutions setting up their Asian headquarters in Hong Kong. Till today, Hong Kong has been ranked as the world's freest economy for many years running, and it remains among one of the world's top 5 leading financial centers.

This period was also marked by military conflict. Wars driven by the Cold War, notably in Vietnam and Afghanistan, wrecked the economies of these respective nations. When the Soviet Union collapsed in 1990-91, many Central Asian states were cut free and were forced to adapt to pressure for democratic and economic change. Also, several of the USSR's allies lost valuable aid and funding.

### *1991-2007*

After the liberalization of the economy of India, undertaken by then finance minister and current Prime Minister of India, Dr. Manmohan Singh, the Indian economy coupled with the Chinese economy to power Asia into being one of the hotspots for world trade. The Chinese economy was already booming under the economic measures undertaken by Deng Xiaoping, in the 1980s, and continuing under Jiang Zemin in the 1990s. In 2007, China's economic growth rate exceeded 11% while India's growth rate increased to around 9%. One of the factors was the sheer size of the population in this region.

Meanwhile, Thailand, Malaysia and Indonesia emerged as the new Asian tigers with their GDPs growing well above 7% per year in the 1980s and the 1990s. Their economies were mainly driven by growing exports. The Philippines only began to open up its stagnated economy in the early 1990s. Vietnam's economy began to grow in 1995, shortly after the United States and Vietnam restored economic and political ties.

Throughout the 1990s, the manufacturing ability and cheap labour markets in Asian developing nations allowed companies to establish themselves in many of the industries previously dominated by companies from developed nations. Asia became one of the largest sources of automobiles, machinery, audio equipment and other electronics.

At the end of 1997, Thailand was hit by currency speculators, and the value of the Baht along with its annual growth rate fell dramatically. Soon after, the crisis spread to Indonesia, Malaysia, South Korea, Hong Kong, Singapore and many other Asian economies, resulting in great economic damage on the affected countries (Japan largely escaped the crisis). In fact, some of the economies, most notably those of Thailand, Indonesia, and South Korea actually contracted. This later would be known as the Asian financial crisis. By 1999, most countries had already recovered from the crisis.

In 2004, parts of Sumatra and South Asia were severely damaged by an earthquake and the subsequent tsunami. The natural disaster wiped out huge amounts of infrastructure throughout the affected area and displaced millions.

### *Future*

Asia's large economic disparities are a source of major continuing tension in the region. While global economic powers Japan, China, India, and South Korea continue powering through, and Indonesia, The Philippines, Malaysia, Thailand, and Vietnam have entered the path to long-term growth, regions right next to these countries are in need of severe assistance.

Given the large number cheap and amply available labour in the region, particularly in China and India, where large workforces provide an economical advantage over other countries, the rising standard of living will eventually lead to a slow-down. Asia is also riddled with political problems that threaten not just the economies, but the general stability of the region and world. The nuclear neighbours—Pakistan and India—constantly pose a threat to each other, causing their governments to heavily invest in military spending.

Military intervention by the United States in Iraq and Afghanistan has also inflamed extremism and resulted in several terrorist attacks in a number of Asian countries. Another impending crisis is the depletion of oil reserves in the Middle East. Most of these economies have traditionally been over-dependent on oil and have had difficulty establishing another pillar in their economies. Yet another potential global danger posed by the economy of Asia is the growing accumulation of foreign exchange reserves. The countries/regions with the largest foreign reserves are mostly in Asia-China (Mainland-$988 billion & Hong Kong-$130 billion, September 2006), Japan ($881 billion, September 2006), Russia ($412 billion, June 2009), India ($277 billion, Mar 2010), Taiwan ($261 billion, September 2006), the Republic of Korea ($228 billion, September 2006), Singapore ($129 billion, June 2005). This increasingly means that the interchangeability of the Euro, USD, and GBP are heavily influenced by Asian central banks. Some economists in the western countries see this as a bad thing, prompting their respective governments to take action.

The economies of Asia are expected to be unequally divided for a long period of time. East Asian nations such as economic leaders Japan, China and South Korea will continue to flourish. Japanese products such as Sony and Hitachi are commanding premium prices in the western world. South Korean conglomerates Samsung and LG are respectively the second and fifth largest in Asia in terms of annual revenues. These two rivalling countries are expected to be joined by new economic competitors such as China and India. On the other hand, the Middle East and a few parts of South East Asia are will be in a state of trouble.

## Economic Sectors

### *Primary Sector*

Asia is by a considerable margin the largest continent in the world, and is rich in natural resources. The vast expanse of the former Soviet Union, particularly that of Russia, contains a huge variety of metals, such as gold, iron, lead, titanium, uranium, and zinc. These metals are mined, but inefficiently due to continued use of poorly maintained, obsolete machinery left over from the communist era. Nevertheless, profits are high due to a commodity price boom in 2003/2004 caused largely by increased demand in China. Oil is Southwest Asia's most important natural resource. Saudi Arabia, Iraq, and Kuwait are rich in oil reserves and have benefited from recent oil price escalations. Asia is home to some four billion people, and thus has a well established tradition in agriculture. High productivity in agriculture, especially of rice, allows high population density of many countries such as Bangladesh, Pakistan, southern China, Cambodia, India, and Vietnam. Agriculture constitutes a high portion of land usage in warm and humid areas of Asia. Many hillsides are farmed in a *terrace* method to boost arable land. The main agricultural products in Asia include rice and wheat. Opium is one of major cash crops in Central and Southeast Asia, particularly in Afghanistan, though its production is prohibited everywhere. Forestry is extensive throughout Asia except Southwest and Central Asia, with many of the items of furniture sold in the developed nations made out of Asian timber. Fishing is a major source of food, particularly in Japan.

### *Secondary Sector*

The manufacturing sector in Asia has traditionally been strongest in the East region-particularly in China, Taiwan, Japan, South Korea and Singapore. The industry varies from manufacturing cheap low value goods such as toys to high-tech added value goods such as computers, CD players, Games consoles, mobile phones and cars. Major Asian manufacturing companies are mostly based in either South Korea or Japan. They include Samsung, Hyundai, LG, and Kia from South Korea, and Sony, Toyota, Toshiba, and Honda from Japan. Many developed-nation firms from Europe, North America, Japan and South Korea have significant operations in the developing Asia to take advantage of the abundant supply of cheap labour. One of the major employers in manufacturing in Asia is the textile industry. Much of the world's supply of clothing and footwear now originates in Southeast Asia and South Asia, particularly in Vietnam, China, India, Thailand, Bangladesh, Pakistan, and Indonesia.

### *Tertiary Sector*

Asia has six important financial centers, located in Dubai, Mumbai, Hong Kong, Shanghai, Singapore and Tokyo. India has been one of the greatest beneficiaries of the economic boom. The country has emerged as one of the world's largest exporters of software and other information technology related services. World class Indian software giants such as Infosys, HCL, Wipro, Mahindra Satyam and TCS have emerged as the world's most sought after service providers. Call centers are also becoming major employers in India and Philippines due to the availability of many well educated English speakers. Here again India holds close to 60% of the trade share. The rise of the Business Process Outsourcing (BPO) industry has seen the rise of India and China as the other financial centers. Experts believe that the current centre of financial activity is moving toward "Chindia"-a name used for jointly referring to China and India-with Shanghai and Mumbai becoming major financial hubs in their own right. Other growing technological and financial hubs include Dhaka (Bangladesh), Bangalore (India), Chennai (India), New Delhi (India), Hyderabad (India), Jakarta (Indonesia), Kuala Lumpur (Malaysia), Kolkata (India), Karachi (Pakistan), Manila (Philippines) and Bangkok (Thailand).

China was the largest and most advanced economy on earth for much of recorded history, until the British Empire (excluding India) overtook it in the mid 19th century. Japan has had for only several decades after WW2 the largest economy in Asia and second-largest of any single nation in the world, after surpassing the Soviet Union (measured in net material product) in 1986 and Germany in 1968. (NB: A number of supernational economies are larger, such as the European Union (EU), the North American Free Trade Agreement (NAFTA) or APEC).

In the late 1980s and early 1990s, Japan's GDP was almost as large (current exchange rate method) as that of the rest of Asia combined. In 1995, Japan's economy nearly equaled that of the USA to tie as the largest economy in the world for a day, after the Japanese currency reached a record high of 79 yen/dollar. Economic growth in Asia since World War II to the 1990s had been concentrated in Japan as well as the four regions of South Korea, Taiwan, Hong Kong and Singapore located in the Pacific Rim, known as the Asian tigers, which have now all received developed country status, having the highest GDP per capita in Asia.

It is forecasted that India will overtake Japan in terms of Nominal GDP by 2020. In terms of GDP per capita, both nominal and PPP-adjusted, South Korea will become the second wealthiest country in Asia by 2025, overtaking Germany, the United Kingdom and France. By 2050, according to a 2006 report by Price Waterhouse Cooper, China will have the largest economy in the world (43% greater than the United States when PPP adjusted, although perhaps smaller than the United States in nominal terms).

Trade blocs:

- Asia-Pacific Economic Cooperation
- Asia-Europe Economic Meeting
- Association of Southeast Asian Nations
- Gulf Cooperation Council
- Closer Economic Partnership Arrangement
- Commonwealth of Independent States
- South Asian Association for Regional Cooperation.

### *Natural Resources*

Asia is the largest continent in the world by a considerable margin, and it is rich in natural resources, such as petroleum, forests, fish, water, rice, copper and silver.

### *Manufacturing*

Manufacturing in Asia has traditionally been strongest in East and Southeast Asia, particularly in mainland China, Taiwan, South Korea, Japan, India, Philippines and Singapore. Japan and South Korea continue to dominate in the area of multinational corporations, but increasingly mainland China, and India are making significant inroads. Many companies from Europe, North America, South Korea and Japan have operations in Asia's developing countries to take advantage of its abundant supply of cheap labour and relatively developed infrastructure.

### *Financial and other Services*

Asia has four main financial centres: Tokyo, Hong Kong, Singapore and Shanghai. Call centres and business process outsourcing (BPOs) are becoming major employers in India and the Philippines due to the availability of a large pool of highly skilled, English-speaking workers. The increased use of outsourcing has assisted the rise of India and the China as financial centres. Due to its large and extremely competitive information technology industry, India has become a major hub for outsourcing.

## Early History

The coastal periphery was home to some of the world's earliest known civilizations, each of them developing around fertile river valleys. The civilizations in Mesopotamia, the Indus Valley and the Huanghe shared many similarities. These civilizations may well have exchanged technologies and ideas such as mathematics and the wheel. Other innovations, such as writing, seem to have been developed individually in each area. Cities, states and empires developed in these lowlands.

The central steppe region had long been inhabited by horse-mounted nomads who could reach all areas of Asia from the steppes. The earliest postulated expansion out of the steppe is that of the Indo-Europeans, who spread their languages into the Middle East, South Asia, and the borders of China, where the Tocharians resided. The northernmost part of Asia, including much of Siberia, was largely inaccessible to the steppe nomads, owing to the dense forests, climate and tundra. These areas remained very sparsely populated.

The centre and the peripheries were mostly kept separated by mountains and deserts. The Caucasus and Himalaya mountains and the Karakum and Gobi deserts formed barriers that the steppe horsemen could cross only with difficulty. While the urban city dwellers were more advanced technologically and socially, in many cases they could do little in a military aspect to defend against the mounted hordes of the steppe. However, the lowlands did not have enough open grasslands to support a large horsebound force; for this and other reasons, the nomads who conquered states in China, India, and the Middle East often found themselves adapting to the local, more affluent societies.

### *Languages and Literature*

Asia is home to several language families and many language isolates. Most Asian countries have more than one language that is natively spoken. For instance, according to Ethnologue, more than 600 languages are spoken in Indonesia, more than 800 languages spoken in India, and more than 100 are spoken in the Philippines. China has many languages and dialects in different provinces.

## *Nobel Prizes*

The polymath Rabindranath Tagore, a Bengali poet, dramatist, and writer from Santiniketan, now in West Bengal, India, became in 1913 the first Asian Nobel laureate. He won his Nobel Prize in Literature for notable impact his prose works and poetic thought had on English, French, and other national literatures of Europe and the Americas. He is also the writer of the national anthems of Bangladesh and India.

Tagore is said to have named another Bengali Indian Nobel prize winner, the 1998 laureate in Economics, Amartya Sen. Sen's work has centered around global issues including famine, welfare, and third-world development. Amartya Sen was Master of Trinity College, Cambridge University, UK, from 1998–2004, becoming the first Asian to head an 'Oxbridge' College.

Other Asian writers who won Nobel Prizes include Yasunari Kawabata (Japan, 1966), Kenzaburô (Japan, 1994), Gao Xingjian (People's Republic of China, 2000) and Orhan Pamuk (Turkey, 2006).

Also, Mother Teresa of India and Shirin Ebadi of Iran were awarded the Nobel Peace Prize for their significant and pioneering efforts for democracy and human rights, especially for the rights of women and children. Ebadi is the first Iranian and the first Muslim woman to receive the prize. Another Nobel Peace Prize winner is Aung San Suu Kyi from Burma for her peaceful and non-violent struggle under a military dictatorship in Burma. She is a nonviolent pro-democracy activist and leader of the National League for Democracy in Burma(Myanmar) and a noted prisoner of conscience. She is a Buddhist and was awarded the Nobel Peace Prize in 1991.

Sir C.V.Raman is the first Asian to get a Nobel prize in Sciences. He won the Nobel Prize in Physics "for his work on the scattering of light and for the discovery of the effect named after him".

Other Asian Nobel Prize winners include Subrahmanyan Chandrasekhar, Abdus Salam, Shmuel Yosef Agnon, Robert Aumann, Menachem Begin, Aaron Ciechanover, Avram Hershko, Daniel Kahneman, Shimon Peres, Yitzhak Rabin, Yaser Arafat, Jose Ramos Horta and Bishop Carlos Filipe Ximenes Belo of Timor Leste, Kim Dae-jung, and thirteen

Japanese scientists. Most of the said awardees are from Japan and Israel except for Chandrasekhar and Raman (India), Salam (Pakistan), Arafat (Palestinian Territories) and Kim (South Korea).

In 2006, Dr. Muhammad Yunus of Bangladesh was awarded the Nobel Peace Prize for the establishment of Grameen Bank, a community development bank that lends money to poor people, especially women in Bangladesh. Dr. Yunus received his Ph.D. in economics from Vanderbilt University, United States. He is internationally known for the concept of micro credit which allows poor and destitutes with little or no collateral to borrow money. The borrowers typically pay back money within the specified period and the incidence of default is very low.

The Dalai Lama has received approximately eighty-four awards over his spiritual and political career. On 22 June 2006, he became one of only four people ever to be recognized with Honorary Citizenship by the Governor General of Canada. On 28 May 2005, he received the Christmas Humphreys Award from the Buddhist Society in the United Kingdom. Most notable was the Nobel Peace Prize, presented in Oslo, Norway on 10 December 1989.

### *Mythology*

Asian mythology is complex and diverse. The story of the Great Flood for example, as presented to Christians in the Old Testament, is first found in Mesopotamian mythology, in the *Epic of Gilgamesh*. Hindu mythology tells about an avatar of the God Vishnu in the form of a fish who warned Manu of a terrible flood. In ancient Chinese mythology, Shan Hai Jing, the Chinese ruler Da Yu, had to spend 10 years to control a deluge which swept out most of ancient China and was aided by the goddess Nuwa who literally fixed the broken sky through which huge rains were pouring.

### *Religions*

Almost all Asian religions have philosophical character and Asian philosophical traditions cover a large spectrum of philosophical thoughts and writings. Indian philosophy includes Hindu philosophy and Buddhist philosophy. They include elements of non-material pursuits, whereas another school of thought from India, Carvaka, preached the enjoyment of material world. Christianity is also present in most Asian countries.

### *Abrahamic*

The Abrahamic religions of Judaism, Christianity, Islam and Bahai Faith originated in West Asia. Judaism, the oldest of the Abrahamic faiths, is practiced primarily in Israel (which has the world's largest Jewish population), though small communities exist in other countries, such as the Bene Israel in India. In the Philippines and East Timor, Roman Catholicism is the predominant religion; it was introduced by the Spaniards and the Portuguese, respectively. In Armenia, Cyprus, Georgia and Russia, Eastern Orthodoxy is the predominant religion. Various Christian denominations have adherents in portions of the Middle East, as well as China and India. The world's largest Muslim community (within the bounds of one nation) is in Indonesia. South Asia (mainly Pakistan, India and Bangladesh) holds 30% of Muslims. There are also significant Muslim populations in China, Iran, Malaysia, southern Philippines (Mindanao), Russia and most of West Asia and Central Asia. The Bahai Faith originated in Asia, in Iran (Persia), and spread from there to the Ottoman Empire, Central Asia, India, and Burma during the lifetime of Bahaullah. Since the middle of the 20th Century, growth has particularly occurred in other Asian countries, because the Bahai Faith's activities in many Muslim countries has been severely suppressed by authorities.

### *Dharmic and Taoist*

The religions of Hinduism, Buddhism, Jainism and Sikhism originated in India, South Asia. In East Asia, particularly in China and Japan, Confucianism, Taoism and Zen Buddhism took shape.

## Asia and its Geography

Asia is the central and eastern part of Eurasia, comprising approximately fifty countries. It has an area, including islands, of roughly 49,694,700 km$^2$. Asia is joined to Africa by the Isthmus of Suez and to Europe by a long border generally following the Ural Mountains.

Medieval Europeans considered Asia as a continent, a distinct landmass. The European concept of the three continents in the Old World goes back to classical antiquity with the etymology of the word rooted in the ancient Near and Middle East. The demarcation between Asia and Africa is the isthmus of Suez and the Red Sea. The boundary between Asia and Europe is commonly believed to run through the Dardanelles, the Sea of Marmara, the Bosporus, the Black Sea, the Caucasus Mountains, the Caspian Sea, the Ural River to its source, and the Ural Mountains to the Kara Sea near Kara, Russia. However, modern discovery of the extent of Africa and Asia made this definition rather anachronistic, especially in the case of Asia, which would have several regions that would be considered distinct landmasses if these criteria were used (for example, South Asia and East Asia).

Geologists and physical geographers no longer consider Europe and Asia to be separate continents. It is either defined in terms of geological landmasses (physical geography) or tectonic plates (geology). In the former case, Europe is a western peninsula of Eurasia or the Africa-Eurasia landmass. In the latter, Europe and Asia are parts of the Eurasian plate, which excludes the Arabian and Indian tectonic plates.

In human geography, there are two schools of thought. One school follows historical convention and treats Europe and Asia as different continents, categorizing Europe, East Asia (the Orient), South Asia (British India), and the Middle East (Arabia and Persia) as specific regions for more detailed analysis. The other schools equate the word "continent" in terms of geographical region when referring to Europe, and use the term "region" to describe Asia in terms of physical geography. It is becoming increasingly common to substitute the term "region" for "continent" because in linguistic terms, "continent" implies a distinct landmass.

There is much confusion in European languages with the term "Asian", which almost always refers to a subcategory of people from Asia rather than referring to "Asian" defined in term of "Asia", because a category implies homogeneity. In American English, Asian refers to East Asians, while in British English, Asian refers to South Asians. Some definitions of Asia exclude Turkey, the Middle East, or Russia. The term is sometimes used more strictly in reference to Asia Pacific, which does not include the Middle East or Russia, but does include islands in the Pacific Ocean — many of which are considered part of Australasia or Oceania. Asia contains the Indian subcontinent, Arabian peninsula, as well as a piece of the North American plate in Siberia.

## Main Geographical Features

The mean elevation of the continent is 950 m (3,117 ft.), the highest of any in the world. The plateau and mountainous areas broadly sweep SW-NW across Asia, climaxing in the high Tibetan Plateau, rising to the highest peaks in the world in the Himalayas. To the northwest lie plains, while to the south lie the geologically distinct areas of the Arabian Peninsula, Indian subcontinent and Malay Peninsula. Large numbers of islands lie southeast of the continent.

Besides its mainland, Asia includes a large number of islands, including some of the world's largest islands, such as Borneo and Sumatra, and some of the world's most populous islands, such as Java and Honshu. Other prominent islands include Bali, Madura and Sulawesi of Indonesia; Hokkaido, Shikoku, Kyushu and Okinawa of Japan; the Andaman and Nicobar of India; Luzon and Mindanao of the Philippines; Ko Pha Ngan and Ko Samui of Thailand; as well as Sri Lanka, Maldives, Singapore, Hong Kong Island and Sakhalin.

The nations of Indonesia, Brunei, East Timor, Singapore, Japan, Philippines, Taiwan, Sri Lanka, Maldives and Cyprus are solely made up of one or more islands, and have no territory on the mainland.

### *Extreme Points*

The following lists the points of Asia that are furthest north, south, east and west. Some of these points are open to debate, as the definition of Asia is varied.

### *Including Islands*

- Northernmost point — Arctic Cape, Komsomolets Island, Severnaya Zemlya, Russia. (81°13' N)
- Southernmost point — Pamana Island, a small island off Rote Island, Indonesia, has been described as Asia's

southernmost point.. However, if the Cocos (Keeling) Islands are included as part of Asia, then South Island (12°04'S) is the southernmost point.

- Westernmost point — Cape Baba, Turkey (26°4'E). This is the westernmost point of the Asian part of Turkey. Although Turkey governs some more westerly islands in the Aegean and Mediterranean Seas, they are considered part of Europe.
- Easternmost point — Big Diomede, Russia (169°0'W) The International Date Line runs between the Russian Big Diomede and the neighbouring U.S.-governed Little Diomede.
- Pole of inaccessibility : Dzoosotoyn Elisen Desert, Xinjiang, China.

### *Mainland*

- Northernmost point — Cape Chelyuskin, Russia (77°43'N)
- Southernmost point — Cape Piai, Malaysia (1°16'N)
- Westernmost point — Cape Baba, Turkey (26°4'E
- Easternmost point — Cape Dezhnev (East Cape), Russia (169°40'W).

### *Geographical Regions*

Asia is a subregion of Eurasia. For further subdivisions based on that term, see North Eurasia and Central Eurasia.

- Central Asia
- Iranian Plateau
- East Asia
- Far East
- North Asia
- South Asia (also Indian subcontinent)
- Southeast Asia
- Southwest Asia (or West Asia).

### *Central Asia*

There is no absolute consensus in the usage of this term. Usually, Central Asia includes:

- The Central Asian republics of Kazakhstan (excluding

its small European territory), Uzbekistan, Tajikistan, Turkmenistan and Kyrgyzstan
- Afghanistan, Mongolia, Iran and the western regions of China are also sometimes included
- Former Soviet states in the Caucasus region.

Central Asia is currently geopolitically important because international disputes and conflicts over oil pipelines, Nagorno-Karabakh, and Chechnya, as well as the presence of U.S. and U.K. military forces in Afghanistan.

### *East Asia*

This area includes:

- Japan
- North and South Korea on the Korean Peninsula.
- China, but sometimes only the eastern regions
- Taiwan
- Mongolia.

More informally, Southeast Asia is included in East Asia on some occasions.

### *North Asia*

This term is rarely used by geographers, but usually it refers to the larger Asian part of Russia, also known as Siberia. Sometimes the northern parts of other Asian nations, such as Kazakhstan or Mongolia, are also included in North Asia. The term "Northern Eurasia" is sometimes used as a political term that includes all the nations of the former U.S.S.R.

### *Southeast Asia*

This region contains the Malay Peninsula, Indochina and islands in the Indian Ocean and Pacific Ocean. The countries it contains are:

- In mainland Southeast Asia, the countries Burma, Thailand, Laos, Cambodia and Vietnam.
- In maritime Southeast Asia, the countries of Malaysia, Brunei, the Philippines, Singapore and Indonesia (some of the Indonesian islands also lie in the Melanesia region of Oceania). East Timor (also Melanesian) is sometimes included too.

Malaysia is divided in two by the South China Sea, and thus has both a mainland and an island part.

### *West Asia*

West Asia is often called the *Middle East* by Europeans and Americans. *Middle East* (to some interpretations) is often used to also refer to some countries in North Africa. West Asia can be further divided into:

- Anatolia (i.e. Asia Minor), constituting the Asian part of Turkey
- The island nation of Cyprus in the Mediterranean Sea
- The Levant or Near East, which includes Syria, Israel, Jordan, Lebanon, Iraq and the Asian portion of Egypt
- The Arabian Peninsula, including Saudi Arabia, United Arab Emirates, Bahrain, Qatar, Oman, Yemen and Kuwait
- The Caucasus region (which straddles both Asia and Europe), namely Transcaucasia, including a small portion of Russia and, arguably, most if not all of Georgia, Armenia, and Azerbaijan
- The Iranian Plateau, including Iran and Afghanistan.

# 2

# India and Southeastern Asia

India, officially the Republic of India is a country in South Asia. It is the seventh-largest country by geographical area, the second-most populous country with 1.18 billion people, and the most populous democracy in the world. Mainland India is bounded by the Indian Ocean on the south, the Arabian Sea on the west, and the Bay of Bengal on the east; and it is bordered by Pakistan to the west; China, Nepal, and Bhutan to the north; and Bangladesh and Burma to the east. India is in the vicinity of Sri Lanka, and the Maldives in the Indian Ocean, and its Andaman and Nicobar Islands are also in the vicinity of the Indonesian island of Sumatra in the Andaman Sea. India has a coastline of 7,517 kilometres (4,700 mi).

Home to the Indus Valley Civilisation and a region of historic trade routes and vast empires, the Indian subcontinent was identified with its commercial and cultural wealth for much of its long history. Four major religions, Hinduism, Buddhism, Jainism and Sikhism originated here, while Zoroastrianism, Judaism, Christianity and Islam arrived in the first millennium CE and shaped the region's diverse culture. Gradually annexed by the British East India Company from the early eighteenth century and colonised by the United Kingdom from the mid-nineteenth century, India became an independent nation in 1947 after a struggle for independence that was marked by widespread non-violent resistance. India is a federal constitutional republic consisting of 28 states and seven union territories with a parliamentary system of democracy. The Indian economy is the world's eleventh largest economy by nominal GDP and the fourth largest by purchasing power parity. Economic reforms since 1991 have transformed it into one of the fastest growing economies in the world. India is a nuclear weapons state and has the third-largest standing army in the world while its military expenditure ranks tenth. India is considered to be a potential superpower, having a rapidly growing economy and growing political clout. India is one of main troop-contributing countries to United Nations peacekeeping operations. It is a member state of the United Nations and members of Commonwealth of Nations, the South Asian Association for Regional Cooperation, G15, G20, Asia Cooperation Dialogue, the Colombo Plan, Indian Ocean Rim Association for Regional Cooperation and the Non-aligned movement, and is one of the BRIC Countries. A pluralistic, multilingual and multiethnic society, India is also home to a diversity of wildlife in a variety of protected habitats.

## Etymology

The name *India* is derived from *Indus*, which is derived from the Old Persian word *Hindu*, from Sanskrit, *Sindhu*, the historic local appellation for the Indus River. The ancient Greeks referred to the Indians as *Indoi*, the people of the Indus. The Constitution of India and common usage in various Indian languages also recognise *Bharat* as an official name of equal status. The name Bharat is derived from the name of the legendary king Bharata in Hindu Mythology. *Hindustan*, originally a Persian word for "Land of the Hindus" referring to northern India, is also occasionally used as a synonym for all of India.

## History

Stone Age rock shelters with paintings at the Bhimbetka rock shelters in Madhya Pradesh are the earliest known traces of human life in India. The first known permanent settlements appeared over 9,000 years ago and gradually developed into the Indus Valley Civilisation, dating back to 3400 BCE in western India. It was followed by the Vedic period, which laid the foundations of Hinduism and other cultural aspects of early Indian society, and ended in the 500s BCE. From around

550 BCE, many independent kingdoms and republics known as the Mahajanapadas were established across the country.

In the third century BCE, most of South Asia was united into the Maurya Empire by Chandragupta Maurya and flourished under Ashoka the Great. From the third century CE, the Gupta dynasty oversaw the period referred to as ancient "India's Golden Age". Empires in Southern India included those of the Chalukyas, the Cholas and the Vijayanagara Empire. Science, technology, engineering, art, logic, language, literature, mathematics, astronomy, religion and philosophy flourished under the patronage of these kings.

Following invasions from Central Asia between the 10th and 12th centuries, much of North India came under the rule of the Delhi Sultanate and later the Mughal Empire. Under the rule of Akbar the Great, India enjoyed much cultural and economic progress as well as religious harmony. Mughal emperors gradually expanded their empires to cover large parts of the subcontinent. However, in North-Eastern India, the dominant power was the Ahom kingdom of Assam, among the few kingdoms to have resisted Mughal subjugation. The first major threat to Mughal imperial power came from a Hindu Rajput king Maha Rana Pratap of Mewar in the 16th century and later from a Hindu state known as the Maratha confederacy, that ruled much of India in the mid-18th century.

From the 16th century, European powers such as Portugal, the Netherlands, France, and Great Britain established trading posts and later took advantage of internal conflicts to establish colonies in the country. By 1856, most of India was under the control of the British East India Company. A year later, a nationwide insurrection of rebelling military units and kingdoms, known as India's First War of Independence or the Sepoy Mutiny, seriously challenged the Company's control but eventually failed. As a result of the instability, India was brought under the direct rule of the British Crown.

In the 20th century, a nationwide struggle for independence was launched by the Indian National Congress and other political organisations. Indian leader Mahatma Gandhi led millions of people in several national campaigns of non-violent civil disobedience.

On 15 August 1947, India gained independence from British rule, but at the same time the Muslim-majority areas were partitioned to form a separate state of Pakistan. On 26 January 1950, India became a republic and a new constitution came into effect. Since independence, India has faced challenges from religious violence, casteism, naxalism, terrorism and regional separatist insurgencies, especially in Jammu and Kashmir and Northeast India. Since the 1990s terrorist attacks have affected many Indian cities. India has unresolved territorial disputes with the People's Republic of China, which, in 1962, escalated into the Sino-Indian War, and with Pakistan, which resulted in wars in 1947, 1965, 1971 and 1999. India is a founding member of the United Nations (as British India) and the Non-Aligned Movement. In 1974, India conducted an underground nuclear test and five more tests in 1998, making India a nuclear state. Beginning in 1991, significant economic reforms have transformed India into one of the fastest-growing economies in the world, increasing its global clout.

## Government

India is federation with a parliamentary form of government, governed under the Constitution of India. It is a constitutional republic and representative democracy, "in which majority rule is tempered by minority rights protected by law." Federalism in India defines the power distribution between the centre and the states. The government is regulated by a checks and balances defined by Indian Constitution, which serves as the country's supreme legal document.

### *Constitution*

The Constitution of India, the longest and the most exhaustive among constitutions of independent nations in the world, came into force on 26 January 1950. The preamble of the constitution defines India as a sovereign, socialist, secular, democratic republic. India has a bicameral parliament operating under a Westminster-style parliamentary system. Its form of government was traditionally described as being 'quasi-federal' with a strong centre and weaker states, but it has grown increasingly federal since the late 1990s as a result of political, economic and social changes.

### *President and Prime Minister*

The President of India is the head of state elected indirectly by an electoral college for a five-year term. The Prime Minister is the head of government and exercises most executive power. Appointed by the President, the Prime Minister is by convention supported by the party or political alliance holding the majority of seats in the lower house of Parliament. The executive branch consists of the President, Vice-President, and the Council of Ministers (the Cabinet being its executive committee) headed by the Prime Minister. Any minister holding a portfolio must be a member of either house of parliament. In the Indian parliamentary system, the executive is subordinate to the legislature, with the Prime Minister and his Council being directly responsible to the lower house of the Parliament.

### *Legislature*

The Legislature of India is the bicameral Parliament, which consists of the upper house called the Rajya Sabha (Council of States) and the lower house called the Lok Sabha (House of People). The Rajya Sabha, a permanent body, has 245 members serving staggered six year terms. Most are elected indirectly by the state and territorial legislatures in proportion to the state's population. 543 of the Lok Sabha's 545 members are directly elected by popular vote to represent individual constituencies for five year terms. The other two members are nominated by the President from the Anglo-Indian community if the President is of the opinion that the community is not adequately represented.

### *Judiciary*

India has a unitary three-tier judiciary, consisting of the Supreme Court, headed by the Chief Justice of India, 21 High Courts, and a large number of trial courts. The Supreme Court has original jurisdiction over cases involving fundamental rights and over disputes between states and the Centre, and appellate jurisdiction over the High Courts. It is judicially independent, and has the power to declare the law and to strike down Union or State laws which contravene the Constitution. The role as the ultimate interpreter of the Constitution is one of the most important functions of the Supreme Court.

## Politics

India is the most populous democracy in the world. It has operated under a multi-party system for most of its history. For most of the years since independence, the federal government has been led by the Indian National Congress (INC). Politics in the states have been dominated by national parties like the INC, the Bharatiya Janata Party (BJP) and various regional parties. From 1950 to 1990, barring two brief periods, the INC enjoyed a parliamentary majority. The INC was out of power between 1977 and 1980, when the Janata Party won the election owing to public discontent with the state of emergency declared by the then Prime Minister Indira Gandhi. In 1989, a Janata Dal-led National Front coalition in alliance with the Left Front coalition won the elections but managed to stay in power for only two years. As the 1991 elections gave no political party a majority, the INC formed a minority government under Prime Minister P.V. Narasimha Rao and was able to complete its five-year term.

The years 1996–1998 were a period of turmoil in the federal government with several short-lived alliances holding sway. The BJP formed a government briefly in 1996, followed by the United Front coalition that excluded both the BJP and the INC. In 1998, the BJP formed the National Democratic Alliance (NDA) with several other parties and became the first non-Congress government to complete a full five-year term.

In the 2004 Indian elections, the INC won the largest number of Lok Sabha seats and formed a government with a coalition called the United Progressive Alliance (UPA), supported by various Left-leaning parties and members opposed to the BJP. The UPA again came into power in the 2009 general election; however, the representation of the Left leaning parties within the coalition has significantly reduced. Manmohan Singh became the first prime minister since Jawaharlal Nehru in 1962 to be re-elected after completing a full five-year term.

## Foreign Relations and Military

Since its independence in 1947, India has maintained cordial relationships with most nations. It took a leading role in the 1950s by advocating the independence of European colonies in Africa and Asia. India is a member of the Commonwealth of

Nations and a founding member of the Non-Aligned Movement. India was involved in two brief military interventions in neighbouring countries – Indian Peace Keeping Force in Sri Lanka and Operation Cactus in Maldives. After the Sino-Indian War and the Indo-Pakistani War of 1965, India's relationship with the Soviet Union warmed and continued to remain so until the end of the Cold War. India has fought two wars with Pakistan over the Kashmir dispute. A third war between India and Pakistan in 1971 resulted in the creation of Bangladesh (then East Pakistan). Additional skirmishes have taken place between the two nations over the Siachen Glacier. In 1999, India and Pakistan fought an undeclared war over Kargil.

In recent years, India has played an influential role in the SAARC and the WTO. India has provided as many as 55,000 Indian military and police personnel to serve in thirty-five UN peacekeeping operations across four continents. India is also an active participant in various mutlilateral forums, particularly the East Asia Summit and the G8+5. Recent overtures by the Indian government have strengthened relations with the United States and China. In the economic sphere, India has close relationships with other developing nations in South America, Asia and Africa.

India maintains the third-largest military force in the world, which consists of the Indian Army, Navy, Air Force and auxiliary forces such as the Paramilitary Forces, the Coast Guard, and the Strategic Forces Command. The official Indian defence budget for 2010 stood at US$31.9 billion (or 2.12% of GDP). According to a 2008 SIPRI report, India's annual military expenditure in terms of PPP stood at US$72.7 billion. The President of India is the supreme commander of the Indian Armed Forces. India maintains close defence cooperation with Russia, Israel and France, who are the chief suppliers of arms. Defence contractors, such as the Defence Research and Development Organisation (DRDO) and Hindustan Aeronautics (HAL), oversee indigenous development of sophisticated arms and military equipment, including ballistic missiles, fighter aircraft and main battle tanks, to reduce India's dependence on foreign imports.

India became a nuclear power in 1974 after conducting an initial nuclear test, known as the Operation Smiling Buddha, and carried out further underground testing in 1998. Despite criticism and military sanctions, India has consistently refused to sign the CTBT and the NPT. India maintains a "no first use" nuclear policy and is developing nuclear triad capability as a part of its "minimum credible deterrence" doctrine. On 10 October 2008, a civilian nuclear agreement between India and the United States was signed, prior to which India received waivers from the IAEA and the NSG which ended restrictions on nuclear technology commerce and recognized India as the world's *de facto* sixth nuclear weapons state.. On 12 March 2010, Russia signed with India a nuclear reactor deal which will build 16 nuclear reactors in India as part of defence and energy deals. On 28 June 2010, Canada signs with India a nuclear co-operation deal to promote and develop co-operation in civilian nuclear energy.

## Geography of India

The geography of India describes the physical features of India, a country in South Asia that lies entirely on the Indian Plate in the northern portion of the Indo-Australian Plate. The country lies to the north of the equator between 8°4' and 37°6' north latitude and 68°7' and 97°25' east longitude. It is the seventh-largest country in the world, with a total land area of 3,287,263 square kilometres (1,269,219 sq mi). India measures 3,214 km (1,997 mi) from north to south and 2,993 km (1,860 mi) from east to west. It has a land frontier of 15,200 km (9,445 mi) and a coastline of 7,517 km (4,671 mi).

India is bounded to the southwest by the Arabian Sea, to the southeast by the Bay of Bengal and the Indian Ocean to the south. Kanyakumari constitutes the southern tip of the Indian peninsula, which narrows before ending in the Indian Ocean. The southernmost part of India is Indira Point in the Andaman and Nicobar Islands. The Maldives, Sri Lanka and Indonesia are island nations to the south of India with Sri Lanka separated from India by a narrow channel of sea formed by Palk Strait and the Gulf of Mannar. The territorial waters of India extend into the sea to a distance of 12 nautical miles (13.8 mi; 22.2 km) measured from the appropriate baseline.

The northern frontiers of India are defined largely by the Himalayan mountain range where its political boundaries with

China, Bhutan, and Nepal lie. Its western borders with Pakistan lie in the Punjab Plain and the Thar desert. In the far northeast, the Chin Hills and Kachin Hills, deeply forested mountainous regions, separate India from Burma while its political border with Bangladesh is defined by the watershed region of the Indo-Gangetic Plain, the Khasi hills and Mizo Hills.

The Ganges is the longest river originating in India and forms the Indo-Gangetic Plain. The Ganges-Brahmaputra system occupies most of northern, central and eastern India, while the Deccan Plateau occupies most of southern India. Along its western frontier is the Thar Desert, which is the seventh-largest desert in the world.

Officially, India's highest point is K2 at 8,611 m (28,251 ft), though it lies in Gilgit-Baltistan, part of the disputed Kashmir region. Kanchenjunga in Sikkim at 8,598 m (28,209 ft) is the highest point within India's current geographic boundaries. Climate across India ranges from equatorial in the far south, to Alpine in the upper reaches of the Himalayas.

## Geological Development

India is entirely contained on the Indian Plate, a major tectonic plate that was formed when it split off from the ancient continent Gondwanaland. About 90 million years ago, during the late Cretaceous Period, the Indian Plate began moving north at about 15 cm/yr (6 in/yr). About 50 to 55 million years ago, in the Eocene epoch of the Cenozoic Era, the plate collided with Asia after covering a distance of 2,000 to 3,000 km (1,243 to 1,864 mi), having moved faster than any other known plate.

In 2007, German geologists determined that the reason the India Plate moved so quickly is that it is only half as thick as the other plates which formerly constituted Gondwanaland. The collision with the Eurasian Plate along the modern border between India and Nepal formed the orogenic belt that created the Tibetan Plateau and the Himalayas. As of 2009, The India Plate is moving northeast at 5 cm/yr (2 in/yr), while the Eurasian Plate is moving north at only 2 cm/yr (0.8 in/yr). India is thus referred to as the "fastest continent." This is causing the Eurasian Plate to deform, and the India Plate to compress at a rate of 4 mm/yr (0.15 in/yr).

## *Political Geography*

India is divided into twenty-eight states (further subdivided into districts) and seven union territories.

States:

1. Andhra Pradesh
2. Arunachal Pradesh
3. Assam
4. Bihar
5. Chhattisgarh
6. Goa
7. Gujarat
8. Haryana
9. Himachal Pradesh
10. Jammu and Kashmir
11. Jharkhand
12. Karnataka
13. Kerala
14. Madhya Pradesh
15. Maharashtra
16. Manipur
17. Meghalaya
18. Mizoram
19. Nagaland
20. Orissa
21. Punjab
22. Rajasthan
23. Sikkim
24. Tamil Nadu
25. Tripura
26. Uttar Pradesh
27. Uttarakhand
28. West Bengal.

## *Union Territories*

A. Andaman and Nicobar Islands

B. Chandigarh
C. Dadra and Nagar Haveli
D. Daman and Diu
E. Lakshadweep
F. National Capital Territory of Delhi
G. Puducherry.

India's borders run a total length of 15,106.70 km (9,387 mi). Its borders with Pakistan and Bangladesh were delineated according to the Radcliffe Line, which was created in 1947 during Partition of India. Its western border with Pakistan extends up to 3,323 km (2,065 mi), dividing the Punjab region and running along the boundaries of the Thar Desert and the Rann of Kutch. Both nations delineated a Line of Control (LoC) to serve as the informal boundary between the Indian and Pakistan-administered areas of Kashmir. According to India's claim, it shares a 106 km (66 mi) border with Afghanistan in northwestern Kashmir, which is under Pakistani control.

India's border with Bangladesh runs 4,096.70 km (2,546 mi). There are 92 enclaves of Bangladesh on Indian soil and 106 enclaves of India are on Bangladeshi soil. The Teen Bigha Corridor is a strip of land formerly belonging to India on the West Bengal–Bangladesh border which has been leased indefinitely to Bangladesh so that it can access its Dehgram–Angalpota enclaves.

The Line of Actual Control (LAC) is the effective border between India and the People's Republic of China. It traverses 4,057 km along the Indian states of Jammu and Kashmir, Uttarakhand, Himachal Pradesh, Sikkim and Arunachal Pradesh. Both nations lay claim to the Aksai Chin region of northeastern Kashmir, which fell into Chinese control during the Sino-Indian War of 1962. The border with Burma (Myanmar) extends up to 1,643 km (1,021 mi) along the southern borders of India's northeastern states. Located amidst the Himalayan range, India's border with Bhutan runs 699 km (434 mi). The border with Nepal runs 1,751 km (1,088 mi) along the foothills of the Himalayas in northern India. The Siliguri Corridor, narrowed sharply by the borders of Bhutan, Nepal and Bangladesh, connects peninsular India with the northeastern states.

## *Physiographic Regions*

India is divided into seven physiographic regions. They are;

1. The northern mountains including the Himalayas, which includes the Kuen Lun and the Karakoram ranges and the northeast mountain ranges.
2. Indo-Gangetic plains
3. Thar Desert
4. Central Highlands and Deccan Plateau
5. East Coast
6. West Coast
7. Bordering seas and islands.

## *Mountains*

A great arc of mountains, consisting of the Himalayas, Hindu Kush, and Patkai ranges define the northern Indian subcontinent. These were formed by the ongoing tectonic collision of the Indian Plate with the Eurasian Plate that started around 50 million years ago. The mountains in these ranges include some of the world's tallest mountains which act as a natural barrier to cold polar winds. They also facilitate the monsoon winds which in turn influence the climate in India. Rivers originating in these mountains, flow through the fertile Indo–Gangetic plains. These mountains are recognised by biogeographers as the boundary between two of the Earth's great ecozones: the temperate Palearctic that covers most of Eurasia and the tropical and subtropical Indomalaya ecozone which includes the Indian subcontinent, Southeast Asia and Indonesia.

India has eight major mountain ranges having peaks of over 1,000 m (3,281 ft):

The Himalayan range is considered as the world's highest mountain range, with its tallest peak Mt. Everest on the Nepal–China border. They form India's northeastern border, separating it from northeastern Asia. They are one of the world's youngest mountain ranges and extend almost uninterrupted for 2,500 km (1,553 mi), covering an area of 500,000 km (193,051 sq mi). The Himalayas extend from Jammu and Kashmir in the west to Arunachal Pradesh in the east. These states along with Himachal Pradesh, Uttarakhand, and Sikkim lie mostly in the

Himalayan region. Numerous Himalayan peaks rise over 7,000 m (22,966 ft) and the snow line ranges between 6,000 m (19,685 ft) in Sikkim to around 3,000 m (9,843 ft) in Kashmir. Kanchenjunga—on the Sikkim–Nepal border—is the highest point in the area administered by India. Most peaks in the Himalayas remain snowbound throughout the year. The Himalayas act as a barrier to the frigid katabatic winds flowing down from Central Asia. Thus, North India is kept warm or only mildly cooled during winter; in summer, the same phenomenon makes India relatively hot.

The Karakoram is situated in the disputed state of Jammu and Kashmir. It has more than sixty peaks above 7,000 m (22,966 ft), including K2, the second highest peak in the world 8,611 m (28,251 ft). K2 is just 237 m (778 ft) smaller than the 8,848 m (29,029 ft) Mount Everest. The range is about 500 km (311 mi) in length and the most heavily glaciated part of the world outside of the polar regions. The Siachen Glacier at 70 km (43 mi) and the Biafo Glacier at 63 km (39 mi) rank as the world's second and third-longest glaciers outside the polar regions. Just to the west of the northwest end of the Karakoram, lies the Hindu Raj range, beyond which is the Hindu Kush range. The southern boundary of the Karakoram is formed by the Gilgit, Indus and Shyok rivers, which separate the range from the northwestern end of the Himalayas.

The Patkai, or Purvanchal, are situated near India's eastern border with Myanmar. They were created by the same tectonic processes which led to the formation of the Himalayas. The physical features of the Patkai mountains are conical peaks, steep slopes and deep valleys. The Patkai ranges are not as rugged or tall as the Himalayas. There are three hill ranges that come under the Patkai: the Patkai–Bum, the Garo–Khasi–Jaintia and the Lushai hills. The Garo–Khasi range lies in Meghalaya. Mawsynram, a village near Cherrapunji lying on the windward side of these hills, has the distinction of being the wettest place in the world, receiving the highest annual rainfall.

The Vindhya range runs across most of central India, extending 1,050 km (652 mi). The average elevation of these hills is 3,000 m (9,843 ft). They are believed to have been formed by the wastes created by the weathering of the ancient Aravali mountains. Geographically, it separates northern India from southern India. The western end of the range lies in eastern Gujarat, near its border with Madhya Pradesh, and runs east and north, almost meeting the Ganges at Mirzapur.

The Satpura Range begins in eastern Gujarat near the Arabian Sea coast and runs east across Maharashtra, Madhya Pradesh and Chhattisgarh. It extends 900 km (559 mi) with many peaks rising above 1,000 m (3,281 ft). It is triangular in shape, with its apex at Ratnapuri and the two sides being parallel to the Tapti and Narmada rivers. It runs parallel to the Vindhya Range, which lies to the north, and these two east-west ranges divide the Indo–Gangetic plain from the Deccan Plateau located north of River Narmada.

The Aravali Range is the oldest mountain range in India, running across Rajasthan from northeast to southwest direction, extending approximately 800 km (497 mi). The northern end of the range continues as isolated hills and rocky ridges into Haryana, ending near Delhi. The highest peak in this range is Guru Shikhar at Mount Abu, rising to 1,722 m (5,650 ft), lying near the border with Gujarat. The Aravali Range is the eroded stub of an ancient fold mountain system. The range rose in a Precambrian event called the Aravali–Delhi orogen. The range joins two of the ancient segments that make up the Indian craton, the Marwar segment to the northwest of the range, and the Bundelkhand segment to the southeast.

The Western Ghats or Sahyadri mountains run along the western edge of India's Deccan Plateau and separate it from a narrow coastal plain along the Arabian Sea. The range runs approximately 1,600 km (994 mi) from south of the Tapti River near the Gujarat–Maharashtra border and across Maharashtra, Goa, Karnataka, Kerala and Tamil Nadu to the southern tip of the Deccan peninsula. The average elevation is around 1,000 m (3,281 ft). Anai Mudi in the Anaimalai Hills 2,695 m (8,842 ft) in Kerala is the highest peak in the Western Ghats.

The Eastern Ghats are a discontinuous range of mountains, which have been eroded and vivisected by the four major rivers of southern India, the Godavari, Mahanadi, Krishna, and Kaveri. These mountains extend from West Bengal to Orissa, Andhra Pradesh and Tamil Nadu, along the coast and parallel to the

Bay of Bengal. Though not as tall as the Western Ghats, some of its peaks are over 1,000 m (3,281 ft) in height. The Nilgiri hills in Tamil Nadu lies at the junction of the Eastern and Western Ghats.

### *Indo-Gangetic Plain*

The Indo-Gangetic plains, also known as the *Great Plains* are large floodplains of the Indus and the Ganga-Brahmaputra river systems. They run parallel to the Himalaya mountains, from Jammu and Kashmir in the west to Assam in the east and draining most of northern and eastern India. The plains encompass an area of 700,000 square kilometers (270,272 sq mi). The major rivers in this region are the Ganges and the Indus along with their tributaries–Beas, Yamuna, Gomti, Ravi, Chambal, Sutlej and Chenab.

The great plains are sometimes classified into four divisions:

- The Bhabar belt — is adjacent to the foothills of the Himalayas and consists of boulders and pebbles which have been carried down by the river streams. As the porosity of this belt is very high, the streams flow underground. The bhabar is generally narrow with its width varying between 7 to 15 km.
- The Terai belt — lies next to the Bhabar region and is composed of newer alluvium. The underground streams reappear in this region. The region is excessively moist and thickly forested. It also receives heavy rainfall throughout the year and is populated with a variety of wildlife.
- The Bangar belt — consists of older alluvium and forms the alluvial terrace of the flood plains. In the Gangetic plains, it has a low upland covered by laterite deposits.
- The Khadar belt — lies in lowland areas after the Bangar belt. It is made up of fresh newer alluvium which is deposited by the rivers flowing down the plain.

The Indo-Gangetic belt is the world's most extensive expanse of uninterrupted alluvium formed by the deposition of silt by the numerous rivers. The plains are flat making it conducive for irrigation through canals. The area is also rich in ground water sources.

The plains are one of the world's most intensely farmed areas. The main crops grown are rice and wheat, which are grown in rotation. Other important crops grown in the region include maize, sugarcane and cotton. The Indo-Gangetic plains rank among the world's most densely populated areas.

### *Thar Desert*

The Thar Desert (also known as the *Great Indian Desert*) is the world's seventh largest desert; it forms a significant portion of western India and covers an area of 238,700 $km^2$ (92,200 $mile^2$). The desert continues into Pakistan as the Cholistan Desert. Most of the Thar Desert is situated in Rajasthan, covering 61% of its geographic area.

About 10 percent of this ecoregion comprises sand dunes, and the remaining 90 percent consist of craggy rock forms, compacted salt-lake bottoms, and interdunal and fixed dune areas. Annual temperatures can range from 0°C in the winter to over 50°C during the summer. Most of the rainfall received in this region is associated with the short July-September southwest monsoon that brings around 100–500 mm of precipitation. Water is scarce and occurs at great depths, ranging from 30 to 120 m below the ground level. Rainfall is precarious and erratic, ranging from below 120 mm (4.72 inches) in the extreme west to 375 mm (14.75 inches) eastward. The soils of the arid region are generally sandy to sandy-loam in texture. The consistency and depth vary as per the topographical features. The low-lying loams are heavier and may have a hard pan of clay, calcium carbonate or gypsum.

### *Highlands*

The Central Highlands comprise of three main plateaus — the Malwa Plateau in the west, the Deccan Plateau in the south (covering most of the Indian peninsula) and the Chota Nagpur Plateau in the east.

The Malwa Plateau is spread across Rajasthan, Madhya Pradesh and Gujarat. The average elevation of the Malwa plateau is 500 metres, and the landscape generally slopes towards the north. Most of the region is drained by the Chambal River and its tributaries; the western part is drained by the upper reaches of the Mahi River. The Deccan Plateau is a large triangular plateau, bounded by the Vindhyas to the north and flanked by the Eastern and Western Ghats. The Deccan covers a total area of 1.9 million km² (735,000 mile²). It is mostly flat, with elevations ranging from 300 to 600 m (1,000 to 2,000 ft). The average elevation of the plateau is 2,000 feet (600 m) above sea level. The surface slopes from 3,000 feet (900 m) in the west to 1,500 feet (450 m) in the east. It slopes gently from west to east and gives rise to several peninsular rivers such as the Godavari, the Krishna, the Kaveri and the Narmada, which drain into the Bay of Bengal. This region is mostly semi-arid as it lies on the leeward side of both Ghats. Much of the Deccan is covered by thorn scrub forest scattered with small regions of deciduous broadleaf forest. Climate in the Deccan ranges from hot summers to mild winters.

The Chota Nagpur Plateau is situated in eastern India, covering much of Jharkhand and adjacent parts of Orissa, Bihar and Chhattisgarh. Its total area is approximately 65,000 km² (25,000 mile²) and is made up of three smaller plateaus — the Ranchi, Hazaribagh, and Kodarma plateaus. The Ranchi plateau is the largest, with an average elevation of 700 m (2,300 ft). Much of the plateau is forested, covered by the Chota Nagpur dry deciduous forests. Vast reserves of metal ores and coal have been found in the Chota Nagpur plateau. The Kathiawar peninsula in western Gujarat is bounded by the Gulf of Kutch and the Gulf of Khambat. The natural vegetation in most of the peninsula is xeric scrub, part of the Northwestern thorn scrub forests ecoregion.

In western India, the Kutch region in Gujarat and Koyna in Maharashtra are classified as a Zone IV region (high risk) for earthquakes. The Kutch city of Bhuj was the epicentre of the 2001 Gujarat earthquake, which claimed the lives of more than 20,000 people and injured 166,836 while destroying or damaging near a million homes. The 1993 Latur earthquake in Maharashtra killed 7,928 people and injured 30,000. Other areas have a moderate to low risk of an earthquake occurring.

### *Coasts*

The Eastern Coastal Plain is a wide stretch of land lying between the Eastern Ghats and the Bay of Bengal. It stretches from Tamil Nadu in the south to West Bengal in the north. The

Mahanadi, Godavari, Kaveri and Krishna rivers drain these plains and their deltas occupy most of the area. The temperature in the coastal regions exceeds 30 °C (86 °F) coupled with high levels of humidity. The region receives both the northeast and southwest monsoon rains. The southwest monsoon splits into two branches, the Bay of Bengal branch and the Arabian Sea branch. The Bay of Bengal branch moves northwards crossing northeast India in early June. The Arabian Sea branch moves northwards and discharges much of its rain on the windward side of Western Ghats. Annual rainfall in this region averages between 1,000 mm (40 in) and 3,000 mm (120 in). The width of the plains varies between 100 and 130 km (62 to 80 miles). The plains are divided into six regions — the Mahanadi delta, the southern Andhra Pradesh plain, the Krishna-Godavari deltas, the Kanyakumari coast, the Coromandel Coast and sandy coastal.

The Western Coastal Plain is a narrow strip of land sandwiched between the Western Ghats and the Arabian Sea, ranging from 50 to 100 km (30 to 60 miles) in width. It extends from Gujarat in the north and extends through Maharashtra, Goa, Karnataka and Kerala. Numerous rivers and backwaters inundate the region. Originating in the Western Ghats, the rivers are fast-flowing and mostly perennial, leading to the formation of estuaries. Major rivers flowing into the sea are the Tapi, Narmada, Mandovi and Zuari. The coast is divided into 3 parts namely, Konkan, which is situated in Maharashtra, Goa and northern parts of Karnataka; the Kanara in Karnataka and the Malabar Coast in Kerala. Vegetation is mostly deciduous, but the Malabar Coast moist forests constitute a unique ecoregion.

## Islands

The Lakshadweep and the Andaman and Nicobar Islands are India's two major island formations which are classified as union territories. The Lakshadweep Islands lie 200 to 300 km (124 to 186 miles) off the coast of Kerala in the Arabian Sea with an area of 32 km$^2$ (11 sq mi). They consist of 12 atolls, 3 reefs and 5 submerged banks, with a total of about 36 islands and islets. The Andaman and Nicobar Islands are located between 6° and 14° North latitude and 92° and 94° East longitude.

They consist of 572 isles, lying in the Bay of Bengal near the Myanmar coast. It is located 1255 km (780 miles) from Kolkata (Calcutta) and 193 km (120 miles) from Cape Negrais in Myanmar. The territory consists of two island groups, the Andaman Islands and the Nicobar Islands. The Andaman Islands consists of 204 small islands with a total length of 352 km (220 miles). India's only active volcano, Barren Island is situated here, having last erupted in May 2005. The Narcondum is a dormant volcano and there is a mud volcano at Baratang. Indira Point, India's southernmost land point is situated in the Nicobar islands, and lies just 189 km (117 miles) from the Indonesian island of Sumatra to the southeast. The highest point is Mount Thullier at 642 m (2,140 ft).

Significant islands just off the Indian coast include Diu, a former Portuguese enclave; Majuli, Asia's largest freshwater island; Elephanta in the Bombay Harbour; and Sriharikota barrier island in Andhra Pradesh. Salsette Island is India's most populous island on which the city of Mumbai (Bombay) is located. Forty-two islands in the Gulf of Kutch constitute the Marine National Park.

### *Water Bodies*

India has around 14,500 km of inland navigable waterways. There are twelve rivers which are classified as major rivers, with the total catchment area exceeding 2,528,000 km$^2$ (976,000 sq mi). All major rivers of India originate from one of the three main watersheds:

1. The Himalaya and the Karakoram ranges
2. Vindhya and Satpura range in central India
3. Sahyadri or Western Ghats in western India.

The Himalayan river networks are snow-fed and have a perennial supply throughout the year. The other two river systems are dependent on the monsoons and shrink into rivulets during the dry season. The Himalayan rivers that flow westward into Pakistan are the Indus, Beas, Chenab, Ravi, Sutlej, and Jhelum.

The Ganga-Brahmaputra-Meghana system has the largest catchment area of 1,100,000 km$^2$ (420,000 sq mi). The Ganga originates from the Gangotri Glacier in Uttarakhand. It flows southeast, draining into the Bay of Bengal. The Yamuna and Gomti rivers also arise in the western Himalayas and join the Ganga in the plains. The Brahmaputra, another tributary of the Ganga, originates in Tibet and enters India through the far-eastern state of Arunachal Pradesh. It proceeds westwards, joining the Ganges in Bangladesh.

The Chambal, another tributary of the Ganga originates from the Vindhya-Satpura watershed. The river flows eastward. Westward-flowing rivers from this watershed are the Narmada and Tapti, which drain into the Arabian Sea in Gujarat. The river network that flows from east to west constitutes 10% of the total outflow.

The Western Ghats are the source of all Deccan rivers, which include the Mahanadi River through the Mahanadi River Delta, Godavari River, Krishna River and Kaveri River, all draining into the Bay of Bengal. These rivers constitute 20% of India's total outflow.

The heavy southwest monsoon rains cause the Brahmaputra and other rivers to distend their banks, often flooding surrounding areas. Though they provide rice paddy farmers with a largely dependable source of natural irrigation and fertilisation, such floods have killed thousands of people and tend to cause displacements of people in such areas.

Major gulfs include the Gulf of Cambay, Gulf of Kutch and the Gulf of Mannar. Straits include the Palk Strait, which separates India from Sri Lanka and the Ten Degree Channel, which separates the Andamans from the Nicobar Islands and the Eight Degree Channel, which separates the Laccadive and Amindivi Islands from the Minicoy Island towards the south. Important capes include the Kanyakumari, the southern tip of mainland India; Indira Point, the southernmost location of India; Rama's Bridge and Point Calimere. While, Arabian Sea lies on the western side of India, Bay of Bengal and Indian Ocean lie towards the eastern and southern side respectively. Smaller seas include the Laccadive Sea and the Andaman Sea. There are four coral reefs in India, located in the Andaman and Nicobar Islands, Gulf of Mannar, Lakshadweep and Gulf of Kutch. Important lakes include Chilka Lake, the country's largest saltwater lake in Orissa; Kolleru Lake in Andhra

Pradesh; Loktak Lake in Manipur, Dal Lake in Kashmir, Sambhar Lake in Rajasthan and the Sasthamkotta Lake in Kerala.

## *Wetlands*

India's wetland ecosystem is widely distributed from the cold and arid located in the Ladakh region of Jammu and Kashmir, and those with the wet and humid climate of peninsular India. Most of the wetlands are directly or indirectly linked to river networks. The Indian government has identified a total of 71 wetlands for conservation and are part of sanctuaries and national parks. Mangrove forests are present all along the Indian coastline in sheltered estuaries, creeks, backwaters, salt marshes and mudflats. The mangrove area covers a total of 4,461 $km^2$ (1,722 sq mi), which comprises 7% of the world's total mangrove cover. Prominent mangrove covers are located in the Andaman and Nicobar Islands, the Sundarbans delta, the Gulf of Kutch and the deltas of the Mahanadi, Godavari and Krishna rivers. Parts of Maharashtra, Karnataka and Kerala also have large mangrove covers.

The Sundarbans delta is home to the largest mangrove forest in the world. It lies at the mouth of the Ganges and spreads across areas of Bangladesh and West Bengal. The Sundarbans is a UNESCO World Heritage Site, but is identified separately as the Sundarbans (Bangladesh) and the Sundarbans National Park (India). The Sundarbans are intersected by a complex network of tidal waterways, mudflats and small islands of salt-tolerant mangrove forests. The area is known for its diverse fauna, being home to a large variety of species of birds, spotted deer, crocodiles and snakes. Its most famous inhabitant is the Bengal Tiger. It is estimated that there are now 400 Bengal tigers and about 30,000 spotted deer in the area.

The Rann of Kutch is a marshy region located in northwestern Gujarat and the bordering Sindh province of Pakistan. It occupies a total area of 27 900 $km^2$ (10,800 $mile^2$). The region was originally a part of the Arabian Sea. Geologic forces such as earthquakes resulted in the damming up of the region, turning it into a large saltwater lagoon. This area gradually filled with silt thus turning it into a seasonal salt marsh. During the monsoons, the area turns into a shallow marsh, often flooding to knee-depth. After the monsoons, the region turns dry and becomes parched.

### *Climate*

Based on the Koppen system, India hosts six major climatic subtypes, ranging from arid desert in the west, alpine tundra and glaciers in the north, and humid tropical regions supporting rainforests in the southwest and the island territories. Many regions have starkly different microclimates. The nation has four seasons: winter (January–February), summer (March–May), a monsoon (rainy) season (June–September) and a post-monsoon period (October–December).

The Himalayas act as a barrier to the frigid katabatic winds flowing down from Central Asia. Thus, North India is kept warm or only mildly cooled during winter; in summer, the same phenomenon makes India relatively hot. Although the Tropic of Cancer—the boundary between the tropics and subtropics—passes through the middle of India, the whole country is considered to be tropical.

Summer lasts between March and June in most parts of India. Temperatures exceed 40 °C (104 °F) during the day. The coastal regions exceed 30 °C (86 °F) coupled with high levels of humidity. In the Thar desert area temperatures can exceed 45 °C (113 °F). The rain-bearing monsoon clouds are attracted to the low-pressure system created by the Thar Desert. The southwest monsoon splits into two arms, the Bay of Bengal arm and the Arabian Sea arm. The Bay of Bengal arm moves northwards crossing northeast India in early June. The Arabian Sea arm moves northwards and deposits much of its rain on the windward side of Western Ghats. Winters in peninsula India see mild to warm days and cool nights. Further north the temperature is cooler. Temperatures in some parts of the Indian plains sometimes fall below freezing. Most of northern India is plagued by fog during this season. The highest temperature recorded in India was 50.6 °C (123.1 °F) in Alwar in 1955. The lowest was "45 °C ("49.0 °F) in Kashmir.

### *Geology*

India's geological features are classified based on their era of formation. The Precambrian formations of Cudappah and

Vindhyan systems are spread out over the eastern and southern states. A small part of this period is spread over western and central India. The Paleozoic formations from the Cambrian, Ordovician, Silurian and Devonian system are found in the Western Himalaya region in Kashmir and Himachal Pradesh. The Mesozoic Deccan Traps formation is seen over most of the northern Deccan; they are believed to be the result of sub-aerial volcanic activity. The Trap soil is black in colour and conducive to agriculture. The Carboniferous system, Permian System and Triassic systems are seen in the western Himalayas. The Jurassic system is seen in the western Himalayas and Rajasthan.

Tertiary imprints are seen in parts of Manipur, Nagaland, Arunachal Pradesh and along the Himalayan belt. The Cretaceous system is seen in central India in the Vindhyas and part of the Indo-Gangetic plains. The Gondowana system is seen in the Narmada River area in the Vindhyas and Satpuras. The Eocene system is seen in the western Himalayas and Assam. Oligocene formations are seen in Kutch and Assam. The Pleistocene system is found over central India. The Andaman and Nicobar Island are thought to have been formed in this era by volcanoes. The Himalayas were formed by the convergence and deformation of the Indo-Australian and Eurasian Plates. Their continued convergence raises the height of the Himalayas by 1 cm each year.

Soils in India can be classified into 8 categories: alluvial, black, red, laterite, forest, arid and desert, saline and alkaline and peaty and organic soils. Alluvial soil constitute the largest soil group in India, constituting 80% of the total land surface. It is derived from the deposition of silt carried by rivers and are found in the Great Northern plains from Punjab to the Assam valley. Alluvial soil are generally fertile but they lack nitrogen and tend to be phosphoric.

Black soil are well developed in the Deccan lava region of Maharashtra, Gujarat, and Madhya Pradesh. These contain high percentage of clay and are moisture retentive. Red soil are found in Tamil Nadu, Karnataka plateau, Andhra plateau, Chota Nagpur plateau and the Aravallis. These are deficient in nitrogen, phosphorus and humus. Laterite soils are formed in tropical regions with heavy rainfall. Heavy rainfall results in leaching out all soluble material of top layer of soil. These are generally found in Western ghats, Eastern ghats and hilly areas of northeastern states that receive heavy rainfall. Forest soils occur on the slopes of mountains and hills in Himalayas, Western Ghats and Eastern Ghats. These generally consist of large amounts of dead leaves and other organic matter called humus.

## *Natural Resources*

Indian coal production is the 3rd highest in the world according to the 2008 Indian Ministry of Mines estimates. Shown above is a coal mine in Jharkhand.

India's total renewable water resources are estimated at 1,907.8 km3/year. Its annual supply of usable and replenshable groundwater amounts to 350 billion cubic metres. Only 35% of groundwater resources are being utilised. About 44 million tonnes of cargo is moved annually through the country's major rivers and waterways. Groundwater supplies 40% of water in India's irrigation canals. 56% of the land is arable and used for agriculture. Black soils are moisture-retentive and are preferred for dry farming and growing cotton, linseed, etc. Forest soils are used for tea and coffee plantations. Red soil have a wide diffusion of iron content.

Most of India's estimated 5.4 billion barrels (860,000,000 $m^3$) in oil reserves are located in the Mumbai High, upper Assam, Cambay, the Krishna-Godavari and Cauvery basins. India possesses about seventeen trillion cubic feet of natural gas in Andhra Pradesh, Gujarat and Orissa. Uranium is mined in Andhra Pradesh. India has 400 medium-to-high enthalpy thermal springs for producing geothermal energy in seven "provinces" — the Himalayas, Sohana, Cambay, the Narmada-Tapti delta, the Godavari delta and the Andaman and Nicobar Islands (specifically the volcanic Barren Island.)

India is the world's biggest producer of mica blocks and mica splittings. India ranks second amongst the world's largest producers of barites and chromites. The Pleistocene system is rich in minerals. India is the third-largest coal producer in the world and ranks fourth in the production of iron ore. It is the fifth-largest producer of bauxite and crude steel, the seventh-largest of manganese ore and the eighth-largest of aluminium.

India has significant sources of titanium ore, diamonds and limestone. India possesses 24% of the world's known and economically-viable thorium, which is mined along shores of Kerala. Gold had been mined in the now-defunct Kolar Gold Fields in Karnataka.

## Flora and Fauna

The Bengal tiger is the national animal of India. India is home to about half of the world's tiger population but the future of the species is threatened by habitat degradation and poaching.

India, which lies within the Indomalaya ecozone, displays significant biodiversity. One of eighteen megadiverse countries, it is home to 7.6% of all mammalian, 12.6% of all avian, 6.2% of all reptilian, 4.4% of all amphibian, 11.7% of all fish, and 6.0% of all flowering plant species. Many ecoregions, such as the *shola* forests, exhibit extremely high rates of endemism; overall, 33% of Indian plant species are endemic.

India's forest cover ranges from the tropical rainforest of the Andaman Islands, Western Ghats, and North-East India to the coniferous forest of the Himalaya. Between these extremes lie the sal-dominated moist deciduous forest of eastern India; the teak-dominated dry deciduous forest of central and southern India; and the babul-dominated thorn forest of the central Deccan and western Gangetic plain. Important Indian trees include the medicinal neem, widely used in rural Indian herbal remedies. The pipal fig tree, shown on the seals of Mohenjo-daro, shaded Gautama Buddha as he sought enlightenment. According to latest report, less than 12% of India's landmass is covered by dense forests.

Many Indian species are descendants of taxa originating in Gondwana, from which the Indian plate separated. Peninsular India's subsequent movement towards, and collision with, the Laurasian landmass set off a mass exchange of species. However, volcanism and climatic changes 20 million years ago caused the extinction of many endemic Indian forms. Soon thereafter, mammals entered India from Asia through two zoo-geographical passes on either side of the emerging Himalaya. Consequently, among Indian species, only 12.6% of mammals and 4.5% of birds are endemic, contrasting with 45.8% of reptiles and 55.8% of amphibians. Notable endemics are the Nilgiri leaf monkey and the brown and carmine Beddome's toad of the Western Ghats. India contains 172, or 2.9%, of IUCN-designated threatened species. These include the Asiatic Lion, the Bengal Tiger, and the Indian white-rumped vulture, which suffered a near-extinction from ingesting the carrion of diclofenac-treated cattle.

In recent decades, human encroachment has posed a threat to India's wildlife; in response, the system of national parks and protected areas, first established in 1935, was substantially expanded. In 1972, India enacted the Wildlife Protection Act and Project Tiger to safeguard crucial habitat; in addition, the Forest Conservation Act was enacted in 1980. Along with more than five hundred wildlife sanctuaries, India hosts thirteen biosphere reserves, four of which are part of the World Network of Biosphere Reserves; twenty-five wetlands are registered under the Ramsar Convention.

## *Economy*

The economy of India is the eleventh largest economy in the world by nominal GDP (just behind Canada, Spain and Brazil) and the fourth largest by purchasing power parity (PPP). Following strong economic reforms from the socialist inspired economy of a post-independence Indian nation, the country began to develop a fast-paced economic growth, as markets opened for international competition and investment, in the 1990s. In the 21st century, India is an emerging economic power with vast human and natural resources, and a huge knowledge base. Economists predict that by 2020, India will be among the leading economies of the world.

India was under social democratic-based policies from 1947 to 1991. The economy was characterised by extensive regulation, protectionism, public ownership, pervasive corruption and slow growth. Since 1991, continuing economic liberalisation has moved the economy towards a market-based system. A revival of economic reforms and better economic policy in 2000s accelerated India's economic growth rate. In recent years, Indian cities have continued to liberalize business regulations. By 2008, India had established itself as the world's second-fastest growing major economy. However, the year 2009 saw a significant slowdown in India's GDP growth rate to 6.8% as well as the return of a large projected fiscal deficit of 6.8% of GDP which would be among the highest in the world.

India's large service industry accounts for 62.5% of the country's Gross Domestic Product (GDP) while the industrial and agricultural sector contribute 20% and 17.5% respectively. Agriculture is the predominant occupation in India, accounting for about 52% of employment. The service sector makes up a further 34%, and industrial sector around 14%. The labour force totals half a billion workers. Major agricultural products include rice, wheat, oilseed, cotton, jute, tea, sugarcane, potatoes, cattle, water buffalo, sheep, goats, poultry and fish. Major industries include telecommunications, textiles, chemicals, food processing, steel, transportation equipment, cement, mining, petroleum, machinery, information technology enabled services and software.

India's per capita income (nominal) is $1,030, ranked 139th in the world, while its per capita (PPP) of US$2,940 is ranked 128th. Previously a closed economy, India's trade has grown fast. India currently accounts for 1.5% of World trade as of 2007 according to the WTO. According to the World Trade Statistics of the WTO in 2006, India's total merchandise trade (counting exports and imports) was valued at $294 billion in 2006 and India's services trade inclusive of export and import was $143 billion. Thus, India's global economic engagement in 2006 covering both merchandise and services trade was of the order of $437 billion, up by a record 72% from a level of $253 billion in 2004. India's trade has reached a still relatively moderate share 24% of GDP in 2006, up from 6% in 1985.

### *History*

India's economic history can be broadly divided into three eras, beginning with the pre-colonial period lasting up to the 18th century. The advent of British colonisation started the colonial period in the early 19th century, which ended with independence in 1947. The third period stretches from independence in 1947 until now.

### *Pre-colonial*

The citizens of the Indus Valley civilisation, a permanent settlement that flourished between 2800 BC and 1800 BC, practiced agriculture, domesticated animals, used uniform weights and measures, made tools and weapons, and traded with other cities. Evidence of well planned streets, a drainage system and water supply reveals their knowledge of urban planning, which included the world's first urban sanitation systems and the existence of a form of municipal government.

The 1872 census revealed that 99.3% of the population of the region constituting present-day India resided in villages, whose economies were largely isolated and self-sustaining, with agriculture the predominant occupation. This satisfied the food requirements of the village and provided raw materials for hand-based industries, such as textiles, food processing and crafts. Although many kingdoms and rulers issued coins, barter was prevalent. Villages paid a portion of their agricultural produce as revenue to the rulers, while its craftsmen received a part of the crops at harvest time for their services.

Religion, especially Hinduism, and the caste and the joint family systems, played an influential role in shaping economic activities. The caste system functioned much like medieval European guilds, ensuring the division of labour, providing for the training of apprentices and, in some cases, allowing manufacturers to achieve narrow specialization. For instance, in certain regions, producing each variety of cloth was the specialty of a particular sub-caste.

Textiles such as muslin, Calicos, shawls, and agricultural products such as pepper, cinnamon, opium and indigo were exported to Europe, the Middle East and South East Asia in return for gold and silver.

Assessment of India's pre-colonial economy is mostly qualitative, owing to the lack of quantitative information. One estimate puts the revenue of Akbar's Mughal Empire in 1600 at £17.5 million, in contrast with the total revenue of Great Britain in 1800, which totalled £16 million. India, by the time of the arrival of the British, was a largely traditional agrarian economy with a dominant subsistence sector dependent on primitive technology. It existed alongside a competitively developed network of commerce, manufacturing and credit. After the decline of the Mughals, western, central and parts of south and north India were integrated and administered by the Maratha Empire. The Maratha Empire's budget in 1740s, at its peak, was Rs. 100 million. After the loss at Panipat, the Maratha Empire disintegrated into confederate states of Gwalior, Baroda, Indore, Jhansi, Nagpur, Pune and Kolhapur. Gwalior state had a budget of Rs. 30M. However, at this time, British East India company entered the Indian political theatre. Until 1857, when India was firmly under the British crown, the country remained in a state of political instability due to internecine wars and conflicts.

### *Colonial*

Company rule in India brought a major change in the taxation environment from revenue taxes to property taxes, resulting in mass impoverishment and destitution of majority of farmers and led to numerous famines. The economic policies of the British Raj effectively bankrupted India's large handicrafts industry and caused a massive drain of India's resources. Indian Nationalists employed the successful Swadeshi movement, as strategy to diminish British economic superiority by boycotting British products and the reviving the market for domestic-made products and production techniques. India had become a strong market for superior finished European goods. This was because of vast gains made by the Industrial revolution in Europe, the effects of which was deprived to Colonial India.

The Nationalists had hoped to revive the domestic industries that were badly effected by polices implemented by British Raj which had made them uncompetitive to British made goods.

An estimate by Cambridge University historian Angus Maddison reveals that "India's share of the world income fell from 22.6% in 1700, comparable to Europe's share of 23.3%, to a low of 3.8% in 1952". It also created an institutional environment that, on paper, guaranteed property rights among the colonizers, encouraged free trade, and created a single currency with fixed exchange rates, standardized weights and measures, capital markets. It also established a well developed system of railways and telegraphs, a civil service that aimed to be free from political interference, a common-law and an adversarial legal system. India's colonisation by the British coincided with major changes in the world economy—industrialisation, and significant growth in production and trade. However, at the end of colonial rule, India inherited an economy that was one of the poorest in the developing world, with industrial development stalled, agriculture unable to feed a rapidly growing population, India had one of the world's lowest life expectancies, and low rates for literacy.

The impact of the British rule on India's economy is a controversial topic. Leaders of the Indian independence movement, and left-nationalist economic historians have blamed colonial rule for the dismal state of India's economy in its aftermath and that financial strength required for Industrial development in Europe was derived from the wealth taken from Colonies in Asia and Africa. At the same time right-wing historians have countered that India's low economic performance was due to various sectors being in a state of growth and decline due to changes brought in by colonialism and a world that was moving towards industrialization and economic integration.

### *Independence to 1991*

Indian economic policy after independence was influenced by the colonial experience (which was seen by Indian leaders as exploitative in nature) and by those leaders' exposure to Fabian socialism. Policy tended towards protectionism, with a strong emphasis on import substitution, industrialization, state intervention in labour and financial markets, a large public sector, business regulation, and central planning. Five-Year Plans of India resembled central planning in the Soviet Union. Steel, mining, machine tools, water, telecommunications, insurance, and electrical plants, among other industries, were effectively nationalized in the mid-1950s. Elaborate licences, regulations and the accompanying red tape, commonly referred to as Licence Raj, were required to set up business in India between 1947 and 1990.

Jawaharlal Nehru, the first prime minister, along with the statistician Prasanta Chandra Mahalanobis, carried on by Indira Gandhi formulated and oversaw economic policy. They expected favourable outcomes from this strategy, because it involved both public and private sectors and was based on direct and indirect state intervention, rather than the more extreme Soviet-style central command system. The policy of concentrating simultaneously on capital-and technology-intensive heavy industry and subsidizing manual, low-skill cottage industries was criticized by economist Milton Friedman, who thought it would waste capital and labour, and retard the development of small manufacturers. The rate from 1947–80 was derisively referred to as the Hindu rate of growth, because of the unfavourable comparison with growth rates in other Asian countries, especially the "East Asian Tigers".

The Rockefeller Foundation's research in high-yielding varieties of seeds, their introduction after 1965 and the increased use of fertilizers and irrigation are known collectively as the Green Revolution in India, which provided the increase in production needed to make India self-sufficient in food grains, thus improving agriculture in India. Famine in India, once accepted as inevitable, has not returned since independence.

### *Since 1991*

In the late 80s, the government led by Rajiv Gandhi eased restrictions on capacity expansion for incumbents, removed price controls and reduced corporate taxes. While this increased the rate of growth, it also led to high fiscal deficits and a worsening current account. The collapse of the Soviet Union, which was India's major trading partner, and the first Gulf War, which caused a spike in oil prices, caused a major balance-of-payments crisis for India, which found itself facing the prospect of defaulting on its loans. India asked for a $1.8 billion bailout loan from IMF, which in return demanded reforms.

In response, Prime Minister Narasimha Rao along with his finance minister and current Prime Minister of India Dr. Manmohan Singh initiated the economic liberalisation of 1991. The reforms did away with the Licence Raj (investment, industrial and import licensing) and ended many public monopolies, allowing automatic approval of foreign direct investment in many sectors. Since then, the overall direction of liberalisation has remained the same, irrespective of the ruling party, although no party has tried to take on powerful lobbies such as the trade unions and farmers, or contentious issues such as reforming labour laws and reducing agricultural subsidies. Since 1990 India has emerged as one of the fastest-growing economies in the developing world; during this period, the economy has grown constantly, but with a few major setbacks. This has been accompanied by increases in life expectancy, literacy rates and food security.

While the credit rating of India was hit by its nuclear tests in 1998, it has been raised to investment level in 2007 by S&P and Moody's. In 2003, Goldman Sachs predicted that India's GDP in current prices will overtake France and Italy by 2020, Germany, UK and Russia by 2025 and Japan by 2035. By 2035, it was projected to be the third largest economy of the world, behind US and China. India is often seen by most economists as a rising economic superpower and is believed to play a major role in the global economy in the 21st century. In 2009 India purchased 200 Tons of Gold for $6.7 billion from IMF as a total role reversal from 1991.

### *Industry and Services*

Industry accounts for 54.6% of the GDP and employ 17% of the total workforce. However, about one-third of the industrial labour force is engaged in simple household manufacturing only. In absolute terms, India is 16th in the world in terms of nominal factory output..

Economic reforms brought foreign competition, led to privatisation of certain public sector industries, opened up sectors hitherto reserved for the public sector and led to an expansion in the production of fast-moving consumer goods. Post-liberalisation, the Indian private sector, which was usually run by oligopolies of old family firms and required political connections to prosper was faced with foreign competition, including the threat of cheaper Chinese imports. It has since handled the change by squeezing costs, revamping management, focusing on designing new products and relying on low labour costs and technology.

Textile manufacturing is the second largest source for employment after agriculture and accounts for 26% of manufacturing output. Ludhiana produces 90% of woolens in India and is also known as the Manchester of India. Tirupur has gained universal recognition as the leading source of hosiery, knitted garments, casual wear and sportswear. Dharavi slum in Mumbai has gained fame for leather products. Tata Motors' Nano attempts to be the world's cheapest car.

India is fifteenth in services output. It provides employment to 23% of work force, and it is growing fast, growth rate 7.5% in 1991–2000 up from 4.5% in 1951–80. It has the largest share in the GDP, accounting for 55% in 2007 up from 15% in 1950.

Business services (information technology, information technology enabled services, business process outsourcing) are among the fastest growing sectors contributing to one third of the total output of services in 2000. The growth in the IT sector is attributed to increased specialization, and an availability of a large pool of low cost, but highly skilled, educated and fluent English-speaking workers, on the supply side, matched on the demand side by an increased demand from foreign consumers interested in India's service exports, or those looking to outsource their operations. The share of India's IT industry to the country's GDP increased from 4.8 % in 2005-06 to 7% in 2008. In 2009, seven Indian firms were listed among the top 15 technology outsourcing companies in the world. In March 2009, annual revenues from outsourcing operations in India amounted to US$60 billion and this is expected to increase to US$225 billion by 2020.

Organized retail such supermarkets accounts for 24% of the market as of 2008. Regulations prevent most foreign investment in retailing. Moreover, over thirty regulations such as "signboard licences" and "anti-hoarding measures" may have to be complied before a store can open doors. There are taxes for moving goods to states, from states, and even within states.

Tourism in India is relatively undeveloped, but growing at double digits. Some hospitals woo medical tourism.

## *Agriculture*

India ranks second worldwide in farm output. Agriculture and allied sectors like forestry, logging and fishing accounted for 16.6% of the GDP in 2007, employed 60% of the total workforce and despite a steady decline of its share in the GDP, is still the largest economic sector and plays a significant role in the overall socio-economic development of India. Yields per unit area of all crops have grown since 1950, due to the special emphasis placed on agriculture in the five-year plans and steady improvements in irrigation, technology, application of modern agricultural practices and provision of agricultural credit and subsidies since Green revolution in India. However, international comparisons reveal the average yield in India is generally 30% to 50% of the highest average yield in the world.

India is the largest producer in the world of milk, cashew nuts, coconuts, tea, ginger, turmeric and black pepper. It also has the world's largest cattle population: 193 million. It is the second largest producer of wheat, rice, sugar, cotton, silk, peanuts and inland fish. It is the third largest producer of tobacco. India is the largest fruit producer, accounting for 10% of the world fruit production. It is the leading producer of bananas, sapotas and mangoes.

India is the second largest producer and the largest consumer of silk in the world, with the majority of the 77 million kg (2005) production taking place in Karnataka State, particularly in Mysore and the North Bangalore regions of Muddenahalli, Kanivenarayanapura, and Doddaballapura, the upcoming sites of a INR 700 million "Silk City".

### *Banking and Finance*

The Indian money market is classified into: the organised sector (comprising private, public and foreign owned commercial banks and cooperative banks, together known as *scheduled banks*); and the unorganised sector (comprising individual or family owned indigenous bankers or money lenders and non-banking financial companies (NBFCs)). The unorganised sector and microcredit are still preferred over traditional banks in rural and sub-urban areas, especially for non-productive purposes, like ceremonies and short duration loans.

Prime Minister Indira Gandhi nationalised 14 banks in 1969, followed by six others in 1980, and made it mandatory for banks to provide 40% of their net credit to priority sectors like agriculture, small-scale industry, retail trade, small businesses, etc. to ensure that the banks fulfill their social and developmental goals. Since then, the number of bank branches has increased from 10,120 in 1969 to 98,910 in 2003 and the population covered by a branch decreased from 63,800 to 15,000 during the same period. The total deposits increased 32.6 times between 1971 to 1991 compared to 7 times between 1951 to 1971. Despite an increase of rural branches, from 1,860 or 22% of the total number of branches in 1969 to 32,270 or 48%, only

32,270 out of 5 lakh (500,000) villages are covered by a scheduled bank. The public sector banks hold over 75% of total assets of the banking industry, with the private and foreign banks holding 18.2% and 6.5% respectively. Since liberalisation, the government has approved significant banking reforms. While some of these relate to nationalised banks (like encouraging mergers, reducing government interference and increasing profitability and competitiveness), other reforms have opened up the banking and insurance sectors to private and foreign players. More than half of personal savings are invested in physical assets such as land, houses, cattle, and gold.

## *Natural Resources*

India's total cultivable area is 1,269,219 km² (56.78% of total land area), which is decreasing due to constant pressure from an ever growing population and increased urbanisation. India has a total water surface area of 314,400 km² and receives an average annual rainfall of 1,100 mm. Irrigation accounts for 92% of the water utilisation, and comprised 380 km² in 1974, and is expected to rise to 1,050 km² by 2025, with the balance accounted for by industrial and domestic consumers. India's inland water resources comprising rivers, canals, ponds and lakes and marine resources comprising the east and west coasts of the Indian ocean and other gulfs and bays provide employment to nearly 6 million people in the fisheries sector. In 2008, India had the world's third largest fishing industry.

India's major mineral resources include coal, iron, manganese, mica, bauxite, titanium, chromite, limestone and thorium. India meets most of its domestic energy demand through its 92 billion tonnes of coal reserves (about 10% of world's coal reserves).

India's huge thorium reserves — about 25% of world's reserves — is expected to fuel the country's ambitious nuclear energy program in the long-run. India's dwindling uranium reserves stagnated the growth of nuclear energy in the country for many years. However, the Indo-US nuclear deal has paved the way for India to import uranium from other countries. India is also believed to be rich in certain renewable sources of energy with significant future potential such as solar, wind and biofuels (jatropha, sugarcane).

### *Petroleum and Natural Gas*

India's oil reserves, found in Bombay High, parts of Gujarat, Rajasthan and eastern Assam, meet 25% of the country's domestic oil demand. India's total proven oil reserves stand at 11 billion barrels, of which Bombay High is believed to hold 6.1 billion barrels and Mangala Area in Rajasthan an additional 3.6 billion barrels.

In 2009, India imported 2.56 million barrels of oil per day, making it one of largest buyers of crude oil in the world. The petroleum industry in India mostly consists of public sector companies such as Oil and Natural Gas Corporation (ONGC), Hindustan Petroleum Corporation Limited (HPCL) and Indian Oil Corporation Limited (IOCL). There are some major private Indian companies in oil sector such as Reliance Industries Limited (RIL) which operates the world's largest oil refining complex.

### *Pharmaceuticals*

India has a self reliant Pharmaceuticals industry. The majority of its medical consumables are produced domestically. Pharmaceutical Industry in India is dotted with companies like Ranbaxy Laboratories, Dr. Reddy's Laboratories, Cipla which have created a niche for themselves at world level.

Today, India is an exporter to countries like the United States and Russia. In terms of the global market, India currently holds a modest 1-2% share, but it has been growing at approximately 10% per year. Indian Pharmaceutical Industry is often compared to Pharmaceutical Industry in the USA.

## External Trade and Investment

### *Global Trade Relations*

India's economy is mostly dependent on its large internal market with external trade accounting for just 20% of the country's GDP. In 2008, India accounted for 1.45% of global merchandise trade and 2.8% of global commercial services export. Until the liberalization of 1991, India was largely and intentionally isolated from the world markets, to protect its economy and to achieve self-reliance. Foreign trade was subject to import tariffs, export taxes and quantitative restrictions, while foreign direct investment (FDI) was restricted by upper-limit equity participation, restrictions on technology transfer, export obligations and government approvals; these approvals were needed for nearly 60% of new FDI in the industrial sector. The restrictions ensured that FDI averaged only around US$200 million annually between 1985 and 1991; a large percentage of the capital flows consisted of foreign aid, commercial borrowing and deposits of non-resident Indians. India's exports were stagnant for the first 15 years after independence, due to the predominance of tea, jute and cotton manufactures, demand for which was generally inelastic. Imports in the same period consisted predominantly of machinery, equipment and raw materials, due to nascent industrialization.

Since liberalization, the value of India's international trade has become more broad-based and has risen to Rs. 63,080,109 crores in 2003–04 from Rs.1,250 crores in 1950–51. India's major trading partners are China, the US, the UAE, the UK, Japan and the EU. The exports during April 2007 were $12.31 billion up by 16% and import were $17.68 billion with an increase of 18.06% over the previous year. In 2006-07, major export commodities included engineering goods, petroleum products, chemicals and pharmaceuticals, gems and jewellery, textiles and garments, agricultural products, iron ore and other minerals. Major import commodities included crude oil and related products, machinery, electronic goods, gold and silver.

India is a founding-member of General Agreement on Tariffs and Trade (GATT) since 1947 and its successor, the WTO. While participating actively in its general council meetings, India has been crucial in voicing the concerns of the developing world. For instance, India has continued its opposition to the inclusion of such matters as labour and environment issues and other *non-tariff barriers* into the WTO policies.

## Balance of Payments

Since independence, India's balance of payments on its current account has been negative. Since liberalisation in the 1990s (precipitated by a balance of payment crisis), India's exports have been consistently rising, covering 80.3% of its imports in 2002–03, up from 66.2% in 1990–91. India's growing oil import bill is seen as the main driver behind the large current account deficit. In 2007-08, India imported 120.1 million tonnes of crude oil, more than 3/4th of the domestic demand, at a cost of $61.72 billion.

Although India is still a net importer, since 1996–97 its overall balance of payments (i.e., including the capital account balance) has been positive, largely on account of increased foreign direct investment and deposits from non-resident Indians; until this time, the overall balance was only occasionally positive on account of external assistance and commercial borrowings. As a result, India's foreign currency reserves stood at $285 billion in 2008.

Due to the global late-2000s recession, both Indian exports and imports declined by 29.2% and 39.2% respectively in June 2009. The steep decline was because countries hit hardest by the global recession, such as United States and members of the European Union, account for more than 60% of Indian exports. However, since the decline in imports was much sharper compared to the decline in exports, India's trade deficit reduced to 252.5 billion rupee.

India's reliance on external assistance and commercial borrowings has decreased since 1991–92, and since 2002–03, it has gradually been repaying these debts. Declining interest rates and reduced borrowings decreased India's debt service ratio to 4.5% in 2007. In India, External Commercial Borrowings (ECBs) are being permitted by the Government for providing an additional source of funds to Indian corporates. The Ministry of Finance monitors and regulates these borrowings (ECBs) through ECB policy guidelines.

### *Foreign Direct Investment in India*

As the fourth-largest economy in the world in PPP terms, India is a preferred destination for foreign direct investments (FDI); India has strengths in telecommunication, information technology and other significant areas such as auto components, chemicals, apparels, pharmaceuticals, and jewellery. Despite a surge in foreign investments, rigid FDI policies resulted in a significant hindrance. However, due to some positive economic reforms aimed at deregulating the economy and stimulating foreign investment, India has positioned itself as one of the front-runners of the rapidly growing Asia Pacific Region. India has a large pool of skilled managerial and technical expertise. The size of the middle-class population stands at 300 million and represents a growing consumer market. The inordinately high investment from Mauritius is due to routing of international funds through the country given significant capital gains tax advantages; double taxation is avoided due to a tax treaty between India and Mauritius, and Mauriitus is a capital gains tax haven, effectively creating a zero-taxation FDI channel.

India's recently liberalized FDI policy (2005) allows up to a 100% FDI stake in ventures. Industrial policy reforms have substantially reduced industrial licensing requirements, removed restrictions on expansion and facilitated easy access to foreign technology and foreign direct investment FDI. The upward moving growth curve of the real-estate sector owes some credit to a booming economy and liberalized FDI regime. In March 2005, the government amended the rules to allow 100 per cent FDI in the construction business. This automatic route has been permitted in townships, housing, built-up infrastructure and construction development projects including housing, commercial premises, hotels, resorts, hospitals, educational institutions, recreational facilities, and city-and regional-level infrastructure.

A number of changes were approved on the FDI policy to remove the caps in most sectors. Fields which require relaxation in FDI restrictions include civil aviation, construction development, industrial parks, petroleum and natural gas, commodity exchanges, credit-information services and mining. But this still leaves an unfinished agenda of permitting greater foreign investment in politically sensitive areas such as insurance and retailing. FDI inflows into India reached a record $19.5 billion in fiscal year 2006-07 (April-March), according to the government's Secretariat for Industrial Assistance. This was more than double the total of US$7.8bn in the previous fiscal year. The FDI inflow for 2007-08 has been reported as $24 billion and for 2008-09, it is expected to be above $35 billion. A critical factor in determining India's continued economic growth and realizing the potential to be an economic superpower is going to depend on how the government can create incentives for FDI flow across a large number of sectors in India.

***Currency***

The Indian rupee is the only legal tender accepted in India. The exchange rate as on 23 March 2010 is 45.40 INR the USD, 61.45 to a EUR, and 68.19 to a GBP. The Indian rupee is accepted as legal tender in the neighbouring Nepal and Bhutan, both of which peg their currency to that of the Indian rupee. The rupee is divided into 100 paise. The highest-denomination banknote is the 1,000 rupee note; the lowest-denomination coin in circulation is the 25 paise coin (it earlier had 1, 2, 5, 10 and 20 paise coins which have been discontinued by the Reserve Bank of India).

The Rupee hit a record low during early 2009 on account of global recession. However, due to a strong domestic market, India managed to bounce back sooner than the western countries. Since September 2009 there has been a constant appreciation in Rupee versus most Tier 1 currencies. On 11 January 2010 Rupee went as high as 45.50 to a United states dollar and on 10 January 2010 as high as Rupee 73.93 to a British Pound. A rising rupee also prompted Government of India to buy 200 tonnes of Gold from IMF.

The RBI, the country's central bank was established on 1 April 1935. It serves as the nation's monetary authority, regulator and supervisor of the financial system, manager of exchange control and as an issuer of currency. The RBI is governed by a central board, headed by a governor who is appointed by the Central government of India.

***Income and Consumption***

As of 2005:

- 85.7% of the population lives on less than $2.50 (PPP) a day, down from 92.5% in 1981. This is much higher than the 80.5% in Sub-Saharan Africa.
- 75.6% of the population lives on less than $2 a day (PPP), which is around 20 rupees or $0.5 a day in nominal terms. It was down from 86.6%, but is still even more than the 73.0% in Sub-Saharan Africa.
- 24.3% of the population earned less than $1 (PPP, around $0.25 in nominal terms) a day in 2005, down from 42.1% in 1981.

- 41.6% of its population is living below the new international poverty line of $1.25 (PPP) per day, down from 59.8% in 1981. The World Bank further estimates that a third of the global poor now reside in India.

Today, more people can afford a bicycle than ever before. Some 40% of Indian households owns a bicycle, with ownership rates ranging from around 30% to 70% at state level. Housing is modest. According to Times of India, "a majority of Indians have per capita space equivalent to or less than a 10 feet x 10 feet room for their living, sleeping, cooking, washing and toilet needs." and "one in every three urban Indians lives in homes too cramped to exceed even the minimum requirements of a prison cell in the US." The average is 103 sq ft (9.6 $m^2$) per person in rural areas and 117 sq ft (10.9 $m^2$) per person in urban areas.

Around half of Indian children are malnourished. The proportion of underweight children is nearly double that of Sub-Saharan Africa. However, India has not had famines since the Green Revolution in the early 1970s. While poverty in India has reduced significantly, official figures estimate that 27.5% of Indians still lived below the national poverty line of $1 (PPP, around 10 rupees in nominal terms) a day in 2004-2005. A 2007 report by the state-run National Commission for Enterprises in the Unorganised Sector (NCEUS) found that 65% of Indians, or 750 million people, lived on less than 20 rupees per day with most working in "informal labour sector with no job or social security, living in abject poverty."

Since the early 1950s, successive governments have implemented various schemes, under planning, to alleviate poverty, that have met with partial success. All these programmes have relied upon the strategies of the *Food for work* programme and *National Rural Employment Programme* of the 1980s, which attempted to use the unemployed to generate productive assets and build rural infrastructure. In August 2005, the Indian parliament passed the *Rural Employment Guarantee Bill*, the largest programme of this type in terms of cost and coverage, which promises 100 days of minimum wage employment to every rural household in all the India's 600 districts. The question of whether economic reforms have reduced poverty or not has fuelled debates without generating any clear cut answers and has also put political pressure on further economic reforms, especially those involving the downsizing of labour and cutting agricultural subsidies.

### *Employment*

Agricultural and allied sectors accounted for about 60% of the total workforce in 2003 same as in 1993–94. While agriculture has faced stagnation in growth, services have seen a steady growth. Of the total workforce, 8% is in the organised sector, two-thirds of which are in the public sector. The NSSO survey estimated that in 1999–2000, 106 million, nearly 10% of the population were unemployed and the overall unemployment rate was 7.3%, with rural areas doing marginally better (7.2%) than urban areas (7.7%). India's labour force is growing by 2.5% annually, but employment only at 2.3% a year.

Official unemployment exceeds 9%. Regulation and other obstacles have discouraged the emergence of formal businesses and jobs. Almost 30% of workers are casual workers who work only when they are able to get jobs and remain unpaid for the rest of the time. Only 10% of the workforce is in regular employment. India's labour regulations are heavy even by developing country standards and analysts have urged the government to abolish them.

Unemployment in India is characterized by chronic or disguised unemployment. Government schemes that target eradication of both poverty and unemployment (which in recent decades has sent millions of poor and unskilled people into urban areas in search of livelihoods) attempt to solve the problem, by providing financial assistance for setting up businesses, skill honing, setting up public sector enterprises, reservations in governments, etc. The decreased role of the public sector after liberalization has further underlined the need for focusing on better education and has also put political pressure on further reforms.

Child labour is a complex problem that is basically rooted in poverty. The Indian government is implementing the world's largest child labour elimination program, with primary education targeted for ~250 million. Numerous non-governmental and voluntary organizations are also involved. Special investigation cells have been set up in states to enforce existing laws banning employment of children (under 14) in hazardous industries. The allocation of the Government of India for the eradication of child labour was $10 million in 1995-96 and $16 million in 1996-97. The allocation for 2007 is $21 million. In 2006, remittances from Indian migrants overseas made up $27 billion or about 3% of India's GDP.

### *Economic Trends*

In the revised 2007 figures, based on increased and sustaining growth, more inflows into foreign direct investment, Goldman Sachs predicts that "from 2007 to 2020, India's GDP per capita in US$ terms will quadruple", and that the Indian economy will surpass the United States (in US$) by 2043. In spite of the high growth rate, the report stated that India would continue to remain a low-income country for decades to come but could be a "motor for the world economy" if it fulfills its growth potential. Goldman Sachs has outlined 10 things that it needs to do in order to achieve its potential and grow 40 times by 2050.

These are;

1. improve governance
2. raise educational achievement
3. increase quality and quantity of universities
4. control inflation
5. introduce a credible fiscal policy
6. liberalize financial markets
7. increase trade with neighbours
8. increase agricultural productivity
9. improve infrastructure and
10. improve environmental quality.

### *Issues*

***Agriculture:*** Slow agricultural growth is a concern for policymakers as some two-thirds of India's people depend on rural employment for a living. Current agricultural practices are neither economically nor environmentally sustainable and India's yields for many agricultural commodities are low. Poorly maintained irrigation systems and almost universal lack of good extension services are among the factors responsible.

Farmers' access to markets is hampered by poor roads, rudimentary market infrastructure, and excessive regulation.

*– World Bank: "India Country Overview 2008"*

The low productivity in India is a result of the following factors:

- According to "India: Priorities for Agriculture and Rural Development" by World Bank, India's large agricultural subsidies are hampering productivity-enhancing investment. Overregulation of agriculture has increased costs, price risks and uncertainty. Government interventions in labour, land, and credit markets are hurting the market. Infrastructure and services are inadequate.
- Illiteracy, slow progress in implementing land reforms and inadequate or inefficient finance and marketing services for farm produce.
- The average size of land holdings is very small (less than 20,000 $m^2$) and is subject to fragmentation, due to land ceiling acts and in some cases, family disputes. Such small holdings are often over-manned, resulting in disguised unemployment and low productivity of labour.
- Adoption of modern agricultural practices and use of technology is inadequate, hampered by ignorance of such practices, high costs and impracticality in the case of small land holdings.
- World Bank says that the allocation of water is inefficient, unsustainable and inequitable. The irrigation infrastructure is deteriorating. Irrigation facilities are inadequate, as revealed by the fact that only 52.6% of the land was irrigated in 2003–04, which result in farmers still being dependent on rainfall, specifically the Monsoon season. A good monsoon results in a robust growth for the economy as a whole, while a poor monsoon leads to a sluggish growth. Farm credit is regulated by NABARD, which is the statutory apex agent for rural development in the subcontinent.

India has many farm insurance companies that insure wheat, fruit, rice and rubber farmers in the event of natural disasters or catastrophic crop failure, under the supervision of the Ministry of Agriculture. One notable company that provides all of these insurance policies is Agriculture Insurance Company of India and it alone insures almost 20 million farmers.

India's population is growing faster than its ability to produce rice and wheat. The most important structural reform for self-sufficiency is the ITC Limited plan to connect 20,000 villages to the Internet by 2013. This will provide farmers with up to date crop prices for the first time, which should minimise losses incurred from neighbouring producers selling early and in turn facilitate investment in rural areas.

## *Corruption*

Corruption has been one of the pervasive problems affecting India. The economic reforms of 1991 reduced the red tape, bureaucracy and the *Licence Raj* that had strangled private enterprise and was blamed by Chakravarthi Rajagopalachari for the corruption and inefficiencies. Yet, a 2005 study by Transparency International (TI) India found that more than half of those surveyed had firsthand experience of paying bribe or peddling influence to get a job done in a public office.

The Right to Information Act (2005) and equivalent acts in the Indian states, that require government officials to furnish information requested by citizens or face punitive action, computerisation of services and various central and state government acts that established vigilance commissions have considerably reduced corruption or at least have opened up avenues to redress grievances. The 2009 report by Transparency International ranks India at 84$^{th}$ place and states that significant improvements were made by India in reducing corruption.

## *Government*

The current government has concluded that most spending fails to reach its intended recipients. Lant Pritchett calls India's public sector "one of the world's top ten biggest problems — of the order of AIDS and climate change". The Economist's 2008 article about the Indian civil service stated that the Indian central government employs around 3 million people, including "vast armies of paper-shuffling peons". At local level, administration can be worse. It is not unheard of that a majority of a state's assembly seats can be held by convicted criminals. One study found that 25% of public sector teachers and 40% of public sector medical workers could not be found at the workplace. India's absence rates are one of the worst in the world.

### *Education*

India has made huge progress in terms of increasing primary education attendance rate and expanding literacy to approximately two thirds of the population. The right to education at elementary level has been made one of the fundamental rights under the Eighty-Sixth Amendment of 2002. However, the literacy rate of 65% is still lower than the worldwide average and the country suffers from a high dropout rate.

### *Infrastructure*

In the past, development of infrastructure was completely in the hands of the public sector and was plagued by corruption, bureaucratic inefficiencies, urban-bias and an inability to scale investment. India's low spending on power, construction, transportation, telecommunications and real estate, at $31 billion or 6% of GDP in 2002 had prevented India from sustaining higher growth rates. This has prompted the government to partially open up infrastructure to the private sector allowing foreign investment which has helped in a sustained growth rate of close to 9% for the past six quarters.

Some 600 million Indians have no mains electricity at all. While 80% of Indian villages have at least an electricity line, just 44% of rural households have access to electricity. According to a sample of 97,882 households in 2002, electricity was the main source of lighting for 53% of rural households compared to 36% in 1993. Some half of the electricity is stolen, compared with 3% in China. The stolen electricity amounts to 1.5% of GDP. Almost all of the electricity in India is produced by the public sector. Power outages are common. Many buy their own power generators to ensure electricity supply. As of 2005 the electricity production was at 661.6 billion kWh with oil production standing at 785,000 bbl/day. In 2007, electricity demand exceeded supply by 15%. Multi Commodity Exchange has tried to get a permit to offer electricity future markets.

Indian Road Network is developing. Trucking goods from Gurgaon to the port in Mumbai can take up to 10 days. India has the world's third largest road network. Container traffic is growing at 15% a year. Some 60% of India's container traffic is handled by the Jawaharlal Nehru Port Trust in Navi Mumbai. Internet use is rare; there were only 7.57 million broadband lines in India in November 2009, however it is still growing at slower rate and is expected to boom after the launch of 3G and wimax services.

Most urban cities have good water supply water 24 hours a day, while some smaller cities face water shortages in summer season. A World Bank report says it is an institutional problem in water agencies, or "how the agency is embedded in the relationships between politics and the citizens who are the consumers."

### *Labour Laws*

India's labour regulations — among the most restrictive and complex in the world —have constrained the growth of the formal manufacturing sector where these laws have their widest application. Better designed labour regulations can attract more labour-intensive investment and create jobs for India's unemployed millions and those trapped in poor quality jobs. Given the country's momentum of growth, the window of opportunity must not be lost for improving the job prospects for the 80 million new entrants who are expected to join the work force over the next decade.

– *World Bank: India Country Overview 2008.*

India's restrictive labour regulations hamper the large-scale creation of formal industrial jobs.

India ranked 133th on the Ease of Doing Business Index 2010, behind countries such as China (89th), Pakistan (85th), and Nigeria (125th). The Constitution provides protection of child labour, slavery, equality of opportunities and forced labour etc. in form of fundamental rights, but the implementation of provisions cited is a big question mark.

### *Economic Disparities*

Lagging states need to bring more jobs to their people by creating an attractive investment destination. Reforming cumbersome regulatory procedures, improving rural connectivity, establishing law and order, creating a stable platform for natural resource investment that balances business interests with social concerns, and providing rural finance are important.

– *World Bank: India Country Overview 2008*

One of the critical problems facing India's economy is the sharp and growing regional variations among India's different states and territories in terms of per capita income, poverty, availability of infrastructure and socio-economic development. Six low-income states-Bihar, Chhattisgarh, Jharkhand, Madhya Pradesh, Orissa and Uttar Pradesh-are home to more than one third of India's population. Between 1999 and 2008, the annualized growth rates for Maharashtra (9%), Gujarat (8.8%), Haryana (8.7%), or Delhi (7.4%) were much higher than for Bihar (5.1%), Uttar Pradesh (4.4%), or Madhya Pradesh (3.5%). However, In 2009-10, Bihar witnessed a growth of about 12.6%, and ended up becoming the 'next big economy in india', followed by Gujarat with a growth of 11.3%. Poverty rates in rural Orissa (43%) and rural Bihar (40%) are some of the worst in the world. On the other hand, rural Haryana (5.7%) and rural Punjab (2.4%) compare well with middle-income countries.

The five-year plans have attempted to reduce regional disparities by encouraging industrial development in the interior regions, but industries still tend to concentrate around urban areas and port cities After liberalization, the more advanced states are better placed to benefit from them, with infrastructure like well developed ports, urbanisation and an educated and skilled workforce which attract manufacturing and service sectors. The union and state governments of backward regions are trying to reduce the disparities by offering tax holidays, cheap land, etc., and focusing more on sectors like tourism, which although being geographically and historically determined, can become a source of growth and is faster to develop than other sectors.

### *Environment and Health*

On Yale and Columbia's Environmental Performance Index, India's score is 21/100 on sanitation, compared with 67/100 for the region and 48/100 for the country income group.

### *Global Financial Crisis*

Among the large economies, India has been one of the least affected by the 2008 global financial crisis and the recession that followed, with the stimulus measures implemented by the country relatively small. Economic growth rates did decline for several quarters, but not as much as in most other countries and a domestic recession was no more than a remote possibility. The relatively better structural balance of the Indian economy, dominated by domestic consumption, deflected the collateral damage from the collapse of the global financial markets and the steep fall in global trade volumes. More importantly, some of the programs initiated by the Indian government even before the crisis inadvertently proved to be the most effective stimulus measures.The most significant of the nationwide programs, which helped sustain domestic demand in India, was a federal initiative to supply temporary employment to the most economically distressed in rural areas. The program offered assured work to adult members in every poor rural family for a minimum of 100 days a year. Though it was stained by poor implementation and allegations of widespread corruption in its early days, the program has since been judged a moderate success, which is significant in a country where government initiatives have poor track record. In early 2008, the federal government also paid off the nearly $15 billion debt owed by more than 36 million small farmers to commercial banks. This initiative was designed to provide financial support to farmers who were worried by low farm produce prices, and constant droughts in several parts of the country. In the short-term, the program added substantially to the cash available to rural consumers, which supported domestic demand.

In 2009, India's nominal GDP stood at US$1.243 trillion, which makes it the eleventh-largest economy in the world. If PPP is taken into account, India's economy is the fourth largest in the world at US$3.561 trillion, corresponding to a per capita income of US$3,100. The country ranks 139th in nominal GDP per capita and 128th in GDP per capita at PPP. With an average annual GDP growth rate of 5.8% for the past two decades, India is one of the fastest growing economies in the world. India has the world's second largest labour force, with 516.3 million people. In terms of output, the agricultural sector accounts for 28% of GDP; the service and industrial sectors make up 54% and 18% respectively. Major agricultural products include rice, wheat, oilseed, cotton, jute, tea, sugarcane, potatoes; cattle, water buffalo, sheep, goats, poultry; fish. Major industries include textiles, telecommunications, chemicals, food processing, steel, transport equipment, cement, mining, petroleum, machinery, software. India's trade has reached a relatively moderate share of 24% of GDP in 2006, up from 6% in 1985. In 2008, India's share of world trade was about 1.68%. Major exports include petroleum products, textile goods, gems and jewelry, software, engineering goods, chemicals, and leather manufactures. Major imports include crude oil, machinery, gems, fertilizer, chemicals.

From the 1950s to the 1980s, India followed socialist-inspired policies. The economy was shackled by extensive regulation, protectionism, and public ownership, leading to pervasive corruption and slow growth. In 1991, the nation liberalised its economy and has since moved towards a market-based system. The policy change in 1991 came after an acute balance of payments crisis, and the emphasis since then has been to use foreign trade and foreign investment as integral parts of India's economy.

In the late 2000s, India's economic growth averaged 7.5% a year. Over the past decade, hourly wage rates in India have more than doubled. In 2009, the Global Competitiveness Report ranked India 16th in financial market sophistication, 24th in banking sector, 27th in business sophistication and 30th in innovation; ahead of several advanced economies. Seven of the world's top 15 technology outsourcing companies are based in India and the country is viewed as the second most favourable outsourcing destination after the United States.

Despite India's impressive economic growth over recent decades, it still contains the largest concentration of poor people in the world. The percentage of people living below the World Bank's international poverty line of $1.25 a day (PPP, in nominal terms Rs. 21.6 a day in urban areas and Rs. 14.3 in rural areas in 2005) decreased from 60% in 1981 to 42% in 2005. Since 1991, inter-state economic inequality in India has consistently grown; the per capita net state domestic product of India's richest states is about 3.2 times that of the poorest states. Even though India has avoided famines in recent decades, half of children are underweight and about 46% of Indian children under the age of three suffer from malnutrition.

A 2007 Goldman Sachs report projected that "from 2007 to 2020, India's GDP per capita will quadruple," and that the Indian GDP will surpass that of the United States before 2050, but India "will remain a low-income country for several decades, with per capita incomes well below its other BRIC peers." Although the Indian economy has grown steadily over the last two decades; its growth has been uneven when comparing different social groups, economic groups, geographic regions, and rural and urban areas. The World Bank suggests that India must continue to focus on public sector reform, infrastructure, agricultural and rural development, removal of labour regulations, improvement in transport, energy security, and health and nutrition.

## *Demographics*

With an estimated population of 1.2 billion, India is the world's second most populous country. The last 50 years have seen a rapid increase in population due to medical advances and massive increase in agricultural productivity due to the "green revolution". India's urban population increased 11-fold during the twentieth century and is increasingly concentrated in large cities. By 2001 there were 35 million-plus population cities in India, with the largest cities, with a population of over 10 million each, being Mumbai, Delhi and Kolkata. However, as of 2001, more than 70% of India's population continues to reside in rural areas.

India is the world's most culturally, linguistically and genetically diverse geographical entity after the African continent. India is home to two major linguistic families: Indo-Aryan (spoken by about 74% of the population) and Dravidian (spoken by about 24%). Other languages spoken in India come from the Austro-Asiatic and Tibeto-Burman linguistic families. Neither the Constitution of India, nor any Indian law defines any *national language*. Hindi, with the largest number of speakers, is the official language of the union. English is used extensively in business and administration and has the status of a 'subsidiary official language;' it is also important in education, especially as a medium of higher education. In addition, every state and union territory has its own official languages, and the constitution also recognises in particular 21 "scheduled languages".

As per the 2001 census, over 800 million Indians (80.5%) were Hindu. Other religious groups include Muslims (13.4%), Christians (2.3%), Sikhs (1.9%), Buddhists (0.8%), Jains (0.4%), Jews, Zoroastrians and Bahais. Tribals constitute 8.1% of the population. India has the third-highest Muslim population in the world and has the highest population of Muslims for a non-Muslim majority country.

India's literacy rate is 64.8% (53.7% for females and 75.3% for males). The state of Kerala has the highest literacy rate at 91% while Bihar has the lowest at 47%. The national human sex ratio is 944 females per 1,000 males. India's median age is 24.9, and the population growth rate of 1.38% per annum; there are 22.01 births per 1,000 people per year. Though India has one of the world's most diverse and modern healthcare systems, the country continues to face several public health-related challenges. According to the World Health Organization, 900,000 Indians die each year from drinking contaminated water and breathing in polluted air. There are about 60 physicians per 100,000 people in India.

## *Culture*

India's culture is marked by a high degree of syncretism and cultural pluralism. India's cultural tradition dates back to 8,000 BCE and has a continuously recorded history for over 2,500 years. With its roots based in the Indus Valley Tradition, the Indian culture took a distinctive shape during the 11th century BCE Vedic age which laid the foundation of Hindu philosophy, mythology, literary tradition and beliefs and practices, such as Dharma, Karma, Yoga and Mokca. It has managed to preserve established traditions while absorbing new customs, traditions, and ideas from invaders and immigrants and spreading its cultural influence to other parts of Asia, mainly South East and East Asia.

Indian religions form one of the most defining aspects of Indian culture. Major dharmic religions which were founded in India include Hinduism, Buddhism and Jainism. Considered to be a successor to the ancient Vedic religion, Hinduism has been shaped by several schools of thoughts such as the *Advaita Vedanta*, the *Yoga Sutras* and the *Bhakti* movement. Buddhism originated in India in 5th century BCE and prominent early Buddhist schools, such as *Theravada* and *Mahayana*, gained dominance during the Maurya Empire. Though Buddhism entered a period of gradual decline in India 5th century CE onwards, it played an influential role in shaping Indian philosophy and thought.

Traditional Indian society is defined by relatively strict social hierarchy. The Indian caste system describes the social stratification and social restrictions in the Indian subcontinent, in which social classes are defined by thousands of endogamous hereditary groups, often termed as *jatis* or castes. Several influential social reform movements, such as the *Bramho Shomaj*, the *Arya Samaja* and the *Ramakrishna* Mission, have played a pivotal role in the emancipation of Dalits (or "untouchables") and other lower-caste communities in India. However, the majority of Dalits continue to live in segregation and are often persecuted and discriminated against.

Traditional Indian family values are highly respected, and multi-generational patriarchal joint families have been the norm, although nuclear family are becoming common in urban areas. An overwhelming majority of Indians have their marriages arranged by their parents and other respected family members, with the consent of the bride and groom. Marriage is thought to be for life, and the divorce rate is extremely low. Child marriage is still a common practice, with half of women in India marrying before the legal age of 18.

Indian cuisine is characterised by a wide variety of regional styles and sophisticated use of herbs and spices. The staple foods in the region are rice (especially in the south and the east) and wheat (predominantly in the north). Spices, such as black pepper which are now consumed world wide, are originally native to the Indian subcontinent. Chili pepper, which was introduced by the Portuguese, is also widely used in Indian cuisine.

Traditional Indian dress varies across the regions in its colours and styles and depends on various factors, including climate. Popular styles of dress include draped garments such as sari for women and dhoti or lungi for men; in addition, stitched clothes such as salwar kameez for women and kurta-pyjama and European-style trousers and shirts for men, are also popular.

Many Indian festivals are religious in origin, although several are celebrated irrespective of caste and creed. Some popular festivals are Diwali, Ganesh Chaturthi, Ugadi, Thai Pongal, Holi, Onam, Vijayadashami, Durga Puja, Eid ul-Fitr, Bakr-Id, Christmas, Buddha Jayanti and Vaisakhi. India has three national holidays which are observed in all states and union territories — Republic Day, Independence Day and Gandhi Jayanthi. Other sets of holidays, varying between nine and twelve, are officially observed in individual states. Religious practices are an integral part of everyday life and are a very public affair. Indian architecture is one area that represents the diversity of Indian culture. Much of it, including notable monuments such as the Taj Mahal and other examples of Mughal architecture and South Indian architecture, comprises a blend of ancient and varied local traditions from several parts of the country and abroad. Vernacular architecture also displays notable regional variation.

Indian music covers a wide range of traditions and regional styles. Classical music largely encompasses the two genres – North Indian Hindustani, South Indian Carnatic traditions and their various offshoots in the form of regional folk music. Regionalised forms of popular music include filmi and folk music; the syncretic tradition of the *bauls* is a well-known form of the latter.

Indian dance too has diverse *folk* and *classical* forms. Among the well-known folk dances are the *bhangra* of the Punjab, the *bihu* of Assam, the *chhau* of West Bengal, Jharkhand, *Sambalpuri* of Orissa, the *ghoomar* of Rajasthan and the *Lawani* of Maharashtra. Eight dance forms, many with narrative forms and mythological elements, have been accorded classical dance status by India's *National Academy of Music, Dance, and Drama*. These are: *Bharatanatyam* of the state of Tamil Nadu, *Kathak* of Uttar Pradesh, *Kathakali* and *Mohiniyattam* of Kerala, *Kuchipudi* of Andhra Pradesh, *Manipuri* of Manipur, *Odissi* of Orissa and the *Sattriya* of Assam. Theatre in India often incorporates music, dance, and improvised or written dialogue.

Often based on Hindu mythology, but also borrowing from medieval romances, and news of social and political events, Indian theatre includes the *Bhavai* of state of Gujarat, the *Jatra* of West Bengal, the *Nautanki* and *Ramlila* of North India, the *tamasha* of Maharashtra, the *Burrakatha* of Andhra Pradesh, the *Terukkuttu* of Tamil Nadu, and the *Yakshagana* of Karnataka. The Indian film industry is the largest in the world. Bollywood, based in Mumbai, makes commercial Hindi films and is the most prolific film industry in the world. Established traditions also exist in Assamese, Bengali, Kannada, Malayalam, Marathi, Tamil, and Telugu language cinemas.

The earliest works of Indian literature were transmitted orally and only later written down. These included works of Sanskrit literature – such as the early Vedas, the epics Mahabharata and Ramayana, the drama *Abhijnanasakuntalam* (The Recognition of Sakuntala), and poetry such as the *Mahakavya* – and the Tamil language *Sangam* literature. Among Indian writers of the modern era active in Indian languages or English, Rabindranath Tagore won the Nobel Prize in 1913. Southeast Asia (or Southeastern Asia) is a subregion of Asia, consisting of the countries that are geographically south of China, east of India and north of Australia. The region lies on the intersection of geological plates, with heavy seismic and volcanic activity. Southeast Asia consists of two geographic regions: the Asian mainland, and island arcs and archipelagoes to the east and southeast. The mainland section consists of Burma (Myanmar), Cambodia, Laos, Thailand, Vietnam and Peninsular Malaysia while the maritime section consists of Brunei, East Malaysia, East Timor, Indonesia, the Philippines, and Singapore.

Austronesian peoples predominate in this region. The major religions are Islam and Buddhism, followed by Christianity. However a wide variety of religions are found throughout the region, including many Hindu and animist-influenced practices.

## Divisions

### *Political*

Definitions of "Southeast Asia" vary, but most definitions include the area represented by the countries:

- Brunei
- Burma (Myanmar)
- Cambodia
- East Timor (Timor-Leste)
- Indonesia
- Laos
- Malaysia
- Philippines
- Singapore
- Thailand
- Vietnam.

All of the above excluding Timor-Leste are members of the Association of Southeast Asian Nations (commonly abbreviated ASEAN.) The area, together with part of South Asia, was widely known as the East Indies or simply the Indies until the twentieth century. Christmas Island and the Cocos (Keeling) Islands are considered part of Southeast Asia though they are governed by Australia. Sovereignty issues exist over some islands in the South China Sea. Papua is politically part of Southeast Asia through Indonesia, although geographically it is often considered as part of Oceania. As of 2009, Papua New Guinea has stated that it might join ASEAN, indicating a possible switch in its geographic locale.

### *Geographical*

The eastern parts of Indonesia and East Timor (east of Wallace Line) are considered to be geographically parts of Oceania.

Southeast Asia is geographically divided into two subregions, namely Mainland Southeast Asia (or Indochina) and Maritime Southeast Asia (or the similarly defined Malay Archipelago) (Indonesian: *Nusantara*).

Mainland Southeast Asia includes:

- Myanmar (Burma)
- Cambodia
- Laos
- Thailand

- Vietnam
- Peninsular Malaysia.

Maritime Southeast Asia includes:

- East Malaysia
- Brunei
- Indonesia
- Philippines
- Singapore
- Timor-Leste.

The Andaman and Nicobar Islands of India are geographically considered part of Southeast Asia. Bangladesh and the Seven Sister States of India are culturally part of Southeast Asia and sometimes considered both South Asian and Southeast Asian. The Seven Sister States of India are also geographically part of southeast Asia. Hainan Island and several other southern Chinese regions such as Yunnan, Guizhou and Guangxi are considered both East Asian and Southeast Asian. The rest of New Guinea is sometimes included so are Palau, Guam, and the Northern Mariana Islands, which were all part of the Spanish East Indies.

## Countries and Territories Data

### *History*

*Homo sapiens* reached the region by around 45,000 years ago. *Homo floresiensis* seems to have shared some islands with modern humans until only 12,000 years ago, when they became extinct. Austronesian people, who form the majority of the modern population in Indonesia and the Philippines, migrated to South East Asia from Taiwan. They arrived in Indonesia around 2000 BCE, and as they spread through the archipelago, confined the native Melanesian peoples to the far eastern regions.

Solheim and others have shown evidence for a *Nusantao* (*Nusantara*) maritime trading network ranging from Vietnam to the rest of the archipelago as early as 5000 BCE to 1 CE. The peoples of Southeast Asia, especially those of Austronesian descent, have been seafarers for thousands of years, some reaching the island of Madagascar. Their vessels, such as the vinta, were ocean-worthy. Magellan's voyage records how much more maneuvreable their vessels were, as compared to the European ships.

Passage through the Indian Ocean aided the colonization of Madagascar by the Austronesian people, as well as commerce between West Asia and Southeast Asia. Gold from Sumatra is thought to have reached as far west as Rome, while slaves from the Sulu Sea was believed to have been used in Magellan's voyage as a translator.

Originally most people were animist. This was later replaced by Brahmanic Hinduism. Theravada Buddhism soon followed in 525. In 1400s, Islamic influences began to enter. This forced the last Hindu court in Indonesia to retreat to Bali.

In Mainland Southeast Asia, Myanmar, Cambodia and Thailand retained the Theravada form of Buddhism, brought to them from Sri Lanka. This type of Buddhism was fused with the Hindu-influenced Khmer culture.

### *Indianized Kingdoms*

Very little is known about Southeast Asian religious beliefs and practices before the advent of Indian merchants and religious influences from the second century BCE onwards. Prior to the 13th century, Hinduism and Buddhism were the main religions in Southeast Asia.

The Jawa Dwipa Hindu kingdom in Java and Sumatra existed around 200 BCE. The history of the Malay-speaking world begins with the advent of Indian influence, which dates back to at least the 3rd century BC. Indian traders came to the archipelago both for its abundant forest and maritime products and to trade with merchants from China, who also discovered the Malay world at an early date. Both Hinduism and Buddhism were well established in the Malay Peninsula by the beginning of the 1st century CE, and from there spread across the archipelago.

Cambodia was first influenced by Hinduism during the beginning of the Funan kingdom. Hinduism was one of the Khmer Empire's official religions. Cambodia is the home to one of the only two temples dedicated to Brahma in the world. Angkor Wat is also a famous Hindu temple of Cambodia. The

Champa civilization was located in what is today central Vietnam, and was a highly indianized Hindu Kingdom.

The Majapahit Empire was an Indianized kingdom based in eastern Java from 1293 to around 1500. Its greatest ruler was Hayam Wuruk, whose reign from 1350 to 1389 marked the empire's peak when it dominated other kingdoms in the southern Malay Peninsula, Borneo, Sumatra, and Bali. Various sources such as the Nagarakertagama also mention that its influence spanned over parts of Celebes, the Moluccas islands, and some areas of western Papua, making it the largest empire to ever exist in Southeast Asian history.

The Cholas excelled in maritime activity in both military and the mercantile fields. Their raids of Kedah and the Srivijaya, and their continued commercial contacts with the Chinese Empire, enabled them to influence the local cultures. Many of the surviving examples of the Hindu cultural influence found today throughout the Southeast Asia are the result of the Chola expeditions.

### *Islamization of Southeast Asia*

In the 11th century, a turbulent period occurred in the history of Maritime Southeast Asia, the Indian Chola navy crossed the ocean and attacked the Srivijaya kingdom of Sangrama Vijayatungavarman in Kadaram (Kedah), the capital of the powerful maritime kingdom was sacked and the king was taken captive. Along with Kadaram, Pannai in present day Sumatra and Malaiyur and the Malayan peninsula were attacked too. Soon after that, the king of Kedah Phra Ong Mahawangsa became the first ruler to abandon the traditional Hindu faith, and converted to Islam with the Sultanate of Kedah established in year 1136. Samudera Pasai converted to Islam in the year 1267, the King of Malacca Parameswara married with princess of Pasai, the son became the first sultan of Malacca, soon Malacca became the centre of Islam study and maritime trade, other rulers followed suit. Indonesian religious leader and Islamic scholar Hamka (1908–1981) wrote in 1961: *"The development of Islam in Indonesia and Malaya is intimately related to a Chinese Muslim, Admiral Zheng He."*

There are several theories to the Islamization process in Southeast Asia. The first theory is trade. The expansion of trade among West Asia, India and Southeast Asia helped the spread of the religion as Muslim traders brought Islam to the region. The second theory is the role of missionaries or Sufis. The Sufi missionaries played a significant role in spreading the faith by syncretising Islamic ideas with existing local beliefs and religious notions. Finally, the ruling classes embraced Islam and that further aided the permeation of the religion throughout the region. The ruler of the region's most important port, Malacca Sultanate, embraced Islam in the 15th century, heralding a period of accelerated conversion of Islam throughout the region as the religion provided a unifying force among the ruling and trading classes.

## Trade and Colonization

### *China*

Chinese merchants have traded with the region for a long time as evidence of Magellan's voyage records that Brunei possessed more cannon than the European ships so it appears that the Chinese fortified them.

Malaysian legend has it that a Chinese Ming emperor sent a princess, Han Li Po, to Malacca, with a retinue of 500, to marry Sultan Mansur Shah after the emperor was impressed by the wisdom of the sultan. Han Li Po's well (constructed 1459) is now a tourist attraction there, as is Bukit Cina, where her retinue settled.

The strategic value of the Strait of Malacca, which was controlled by Sultanate of Malacca in the 15th and early 16th century, did not go unnoticed by Portuguese writer Duarte Barbosa, who in 1500 wrote "He who is lord of Malacca has his hand on the throat of Venice".

### *Europe*

Western influence started to enter in the 1500s, with the arrival of the Portuguese and Spanish in Moluccas and the Philippines. Later the Dutch established the Dutch East Indies; the French Indochina; and the British Strait Settlements. All southeast Asian countries were colonized except for Thailand.

European explorers were reaching Southeast Asia from the west and from the east. Regular trade between the ships sailing east from the Indian Ocean and south from mainland Asia provided goods in return for natural products, such as honey and hornbill beaks from the islands of the archipelago.

Europeans brought Christianity allowing Christian missionaries to become widespread. Thailand also allowed Western science and technology to enter its country.

### *Japan*

During World War II, the Imperial Japan invaded most of the former western colonies. The Showa occupation regime committed violent actions against indigenous civilians such as the Manila Massacre and the implementation of a system of forced labour, such as the one involving 4 to 10 million *romusha* in Indonesia. A later UN report stated that four million people died in Indonesia as a result of famine and forced labour during the Japanese occupation.

### *Present*

Most countries in the region enjoy national autonomy. Democratic forms of government and the recognition of human rights are taking root. ASEAN provides a framework for the integration of commerce.

Conflicting territorial and maritime claims continue to exist, including the conflicting claims by Taiwan, China, and the Philippines over the Spratly Islands.

### *Geography*

Geologically, the Indonesian archipelago is one of the most active vulcanological regions in the world. Geological uplifts in the region have also produced some impressive mountains, culminating in Puncak Jaya in Papua, Indonesia at 5,030 metres (16,024 ft), on the island of New Guinea, it is the only place where ice glacier can be found in Southeast Asia. While the second tallest peak is Mount Kinabalu in Sabah, Malaysia on the island of Borneo with a height of 4,101 meters (13,455 ft). The tallest mountain in the Southeast Asia is Hkakabo Razi at 5,967 meters and can be found in northern Myanmar. The largest archipelago in the world by size is Indonesia.

### *Boundaries*

The Australian continent defines a region adjacent to

Southeast Asia, which is also politically separated from the countries of Southeast Asia. But a cultural touch point lies between Papua New Guinea and the Indonesian region of Papua and West Papua, which shares the island of New Guinea with Papua New Guinea.

### *Climate*

The climate in Southeast Asia is mainly tropical–hot and humid all year round with plentiful rainfall. Southeast Asia has a wet and dry season caused by seasonal shift in winds or monsoon. The tropical rain belt causes additional rainfall during the monsoon season. The rain forest is the second largest on earth (with the Amazon being the largest). An exception to this type of climate and vegetation is the mountain areas in the northern region, where high altitudes lead to milder temperatures and drier landscape. Other parts fall out of this climate because they are desert like.

### *Environment*

All of Southeast Asia falls within the warm, humid tropics, and its climate generally can be characterized as monsoonal. The animals of Southeast Asia are diverse; on the islands of Borneo and Sumatra, the Orangutan (man of the forest), the Asian Elephant, the Malayan tapir, the Sumatran Rhinoceros and the Bornean Clouded Leopard can be also found. Six subspecies of the Binturong or *bearcat* exist in the region, though the one endemic to the island of Palawan is now classed as vulnerable.

The Komodo Dragon is the largest living species of lizard and inhabits the islands of Komodo, Rinca, Flores, and Gili Motang in Indonesia.

The Wild Asian Water Buffalo, and on various islands related dwarf species of Bubalus such as Anoa were once widespread in Southeast Asia, nowadays the Domestic Asian Water buffalo is common across the region, but its remaining relatives are rare and endangered.

The mouse deer, a small tusked deer as large as a toy dog or cat, can be found on Sumatra, Borneo and Palawan Islands. The gaur, a gigantic wild ox larger than even wild Water buffalo, is found mainly in Indochina.

Birds such as the peafowl and drongo live in this subregion as far east as Indonesia. The babirusa, a four-tusked pig, can be found in Indonesia as well. The hornbill was prized for its beak and used in trade with China. The horn of the rhinoceros, not part of its skull, was prized in China as well.

The Indonesian Archipelago is split by the Wallace Line. This line runs along what is now known to be a tectonic plate boundary, and separates Asian (Western) species from Australasian (Eastern) species. The islands between Java/Borneo and Papua form a mixed zone, where both types occur, known as Wallacea. As the pace of development accelerates and populations continue to expand in Southeast Asia, concern has increased regarding the impact of human activity on the region's environment. A significant portion of Southeast Asia, however, has not changed greatly and remains an unaltered home to wildlife. The nations of the region, with only few exceptions, have become aware of the need to maintain forest cover not only to prevent soil erosion but to preserve the diversity of flora and fauna. Indonesia, for example, has created an extensive system of national parks and preserves for this purpose. Even so, such species as the Javan rhinoceros face extinction, with only a handful of the animals remaining in western Java.

The shallow waters of the Southeast Asian coral reefs have the highest levels of biodiversity for the world's marine ecosystems, where coral, fish and molluscs abound. According to Conservation International, marine surveys suggest that the marine life diversity in the Raja Ampat area is the highest recorded on Earth. Diversity is considerably greater than any other area sampled in the Coral Triangle composed of Indonesia, Philippines and Papua New Guinea. The Coral Triangle is the heart of the world's coral reef biodiversity, making Raja Ampat quite possibly the richest coral reef ecosystems in the world. The whale shark, the world's largest species of fish and 6 species of pawikans can also be found in the South China Sea and the Pacific Ocean territories of the Philippines.

The trees and other plants of the region are tropical; in some countries where the mountains are tall enough, temperate-climate vegetation can be found. These rainforest areas are currently being logged-over, especially in Borneo.

While Southeast Asia is rich in flora and fauna, Southeast Asia is facing severe deforestation which causes habitat loss for various endangered species such as orangutan and the Sumatran tiger. Predictions have been made that more than 40% of the animal and plant species in Southeast Asia could be wiped out in the 21st century. At the same time, haze has been a regular occurrence. The two worst regional hazes were in 1997 and 2006 in which multiple countries were covered with thick haze, mostly caused by "slash and burn" activities in Indonesia. In reaction, several countries in Southeast Asia signed the ASEAN Agreement on Transboundary Haze Pollution in order to combat haze pollution.

## *Economy*

Even prior to the penetration of European interests, Southeast Asia was a critical part of the world trading system. The Ryukyu Kingdom often participated in maritime trade in Southeast Asia. A wide range of commodities originated in the region, but especially important were such spices as pepper, ginger, cloves, and nutmeg. The spice trade initially was developed by Indian and Arab merchants, but it also brought Europeans to the region. First Spaniards (Manila galleon) and Portuguese, then the Dutch, and finally the British and French became involved in this enterprise in various countries. The penetration of European commercial interests gradually evolved into annexation of territories, as traders lobbied for an extension of control to protect and expand their activities. As a result, the Dutch moved into Indonesia, the British into Malaya, the French into Indochina and the Spanish into the Philippines

While the region's economy greatly depends on agriculture, manufacturing and services are becoming more important. An emerging market, Indonesia is the largest economy in this region. Newly industrialized countries include the Philippines, Malaysia, and Thailand, while Singapore and Brunei are affluent developed economies. The rest of Southeast Asia is still heavily dependent on agriculture, but Vietnam is notably making steady progress in developing its industrial sectors. The region notably manufactures textiles, electronic high-tech goods such as microprocessors and heavy industrial products such as automobiles. Reserves of oil are also present in the region.

Seventeen telecommunications companies have contracted to build a new submarine cable to connect Southeast Asia to the U.S. This is to avoid disruption of the kind recently caused by the cutting of the undersea cable from Taiwan to the U.S. in a recent earthquake.

Tourism has been a key factor in economic development for many Southeast Asian countries, especially Cambodia. According to UNESCO, "tourism, if correctly conceived, can be a tremendous development tool and an effective means of preserving the cultural diversity of our planet." Since the early 1990s, "even the non-ASEAN nations such as Cambodia, Laos, Vietnam and Myanmar, where the income derived from tourism is low, are attempting to expand their own tourism industries." In 1995, Singapore was the regional leader in tourism receipts relative to GDP at over 8%. By 1998, those receipts had dropped to less than 6% of GDP while Thailand and Lao PDR increased receipts to over 7%. Since 2000, Cambodia has surpassed all other ASEAN countries and generated almost 15% of its GDP from tourism in 2006.

## *Demographics*

Southeast Asia has an area of approximately 4,000,000 km (1.6 million square miles). As of 2004, more than 593 million people lived in the region, more than a fifth of them (125 million) on the Indonesian island of Java, the most densely populated large island in the world. Indonesia is the most populous country with 230 million people and also 4th most populous country in the world. The distribution of the religions and people is diverse in Southeast Asia and varies by country. Some 30 million overseas Chinese also live in Southeast Asia, most prominently in Christmas Island, Malaysia, the Philippines, Singapore, Indonesia and Thailand, and also, as the Hoa, in Vietnam.

## *Ethnic Groups*

According to a recent Stanford genetic study, the Southeast Asian population is far from being homogeneous. Although primarily descendants of Austronesian, Tai, and Mon-Khmer-speaking immigrants who migrated from Southern China during the Bronze Age and Iron Age, there are overlays of Arab, Chinese, Indian, Polynesian and Melanesian genes.

There are also large pockets of intermarriage between indigenous Southeast Asians and those of Chinese descent. They form a substantial part of everyday life in countries such as Vietnam, Singapore, Thailand and the Philippines. Indonesia and Malaysia also has a few mixed Southeast Asian-Chinese populations. On the mainland the Khmer peoples of Cambodia remain as ancestors of earlier Pareoean peoples. Similarly, remnants of the Mon group are found in parts of Myanmar and Thailand; the ethnic mixture there has been produced by overlaying Tibeto-Burman and Tai, Lao, and Shan peoples. The contemporary Vietnamese population originated from the Red River area in the north and may be a mixture of Tai and Malay peoples. Added to these major ethnic groups are such less numerous peoples as the Karens, Chins, and Nagas in Myanmar, who have affinities with other Asiatic peoples. Insular Southeast Asia contains a mixture of descendants of Proto-Malay (Nesiot) and Pareoean peoples who were influenced by Malayo-Polynesian and other groups. In addition, Arabic, Indian, and Chinese influences have affected the ethnic pattern of the islands.

In modern times, the Javanese are the largest ethnic group in Southeast Asia, with more than 86 million people, mostly concentrated in Java, Indonesia. In Myanmar, the Burmese account for more than two-thirds of the ethnic stock in this country, while ethnic Thais and Vietnamese account for about four-fifths of the respective populations of those countries. Indonesia is clearly dominated by the Javanese and Sundanese ethnic groups, while Malaysia is more evenly split between the Malays and the Chinese. Within the Philippines, the Tagalog, Cebuano, Ilocano, and Bicol groups are significant.

### *Religions*

Islam is the most widely practiced religion in Southeast Asia, numbering approximately 240 million adherents which translate to about 40% of the entire population, with majorities in Brunei, Indonesia and Malaysia. Countries in Southeast Asia practice many different religions. *Mainland* Southeast Asian countries, which are, Thailand, Cambodia, Laos, Myanmar, Vietnam and Singapore practice predominantly Buddhism. Ancestor worship and Confucianism is also widely practiced in Vietnam and Singapore. In Maritime Southeast Asia, people living in Malaysia, western Indonesia and Brunei practice mainly Islam. Christianity is predominant in the Philippines because of Spanish colonization for more than 300 years, eastern Indonesia and East Timor. The Philippines has the largest Roman Catholic population followed very distantly by Vietnam. East Timor is also predominantly Roman Catholic due to a history of Portuguese rule.

The religious composition for each country is as follows. Some values are taken from the *CIA World Factbook*: Religions and peoples are diverse in Southeast Asia and not one country is homogeneous. In the world's most populous Muslim nation, Indonesia, Hinduism is dominant on islands such as Bali. Christianity also predominates in Philippines, New Guinea and Timor.

The phoenix who is the mount (vahanam) of Vishnu, is a national symbol in both Thailand and Indonesia; in the Philippines, gold images of Garuda have been found on Palawan; gold images of other Hindu gods and goddesses have also been found on Mindanao. Balinese Hinduism is somewhat different from Hinduism practiced elsewhere, as Animism and local culture is incorporated into it. Cambodia still involves today in practice Hindism for mixture toward the excessity Buddhism disciple of the numerous special occasion, worship and belief with the root of previous powerful Khmer Empire. Christians can also be found throughout Southeast Asia; they are in the majority in East Timor and the Philippines, Asia's largest Christian nation. In addition, there are also older tribal religious practices in remote areas of Sarawak in East Malaysia and Papua in eastern Indonesia. In Myanmar, Sakka (Indra) is revered as a *nat*. In Vietnam, Mahayana Buddhism is practiced, which is influenced by native animism but with strong emphasis on Ancestor Worship.

## *Languages*

Each of the languages have been influenced by cultural pressures due to trade and historical colonization as well. Thus, for example, a Filipino, educated in English and Filipino, as well as in his native tongue (e.g., Visayan), might well speak another language, such as Spanish for historical reasons, or

Chinese, Korean or Japanese for economic reasons; a Malaysian might well speak English, Chinese, Tamil as well as Malay as a second language.

The language composition for each country is as follows: (official languages are in bold.)

| | |
|---|---|
| Andaman and Nicobar Islands | Nicobarese, Bengali, English, Hindi, Malayalam, Punjabi, Tamil, Telugu, Shompen, Andamanese languages, others |
| Brunei | Malay, English, Chinese, indigenous Borneian dialects Cambodia Khmer, English, French, Vietnamese, Thai, Chamic dialects, Chinese languages, others Christmas Island English, Chinese, Malay Cocos (Keeling) Islands English, Cocos Malay East Timor Tetum, Portuguese, Indonesian, English, Mambae, Makasae, Tukudede, Bunak, Galoli, Kemak, Fataluku, Baikeno, others Indonesia Indonesian, Acehnese, Batak, Minang, Sundanese, Javanese, Banjarese, Sasak, Tetum, Dayak, Minahasa, Toraja, Buginese, Halmahera, Ambonese, Ceramese; Dutch, Papuan languages, Chinese, others Laos Lao, Thai, Vietnamese, Hmong, Miao, Mien, Dao, Shan; French, English others Malaysia Malay, English, Mandarin, Chinese dialects, Indian languages, Sarawakian and Sabahan languages, others Myanmar (Burma) Burmese, Shan, Karen, Rakhine, Kachin, Chin, Mon, Chinese languages, Indian languages, others |
| Philippines | Filipino, English, Tagalog, Cebuano, Ilocano, Hiligaynon/Ilonggo, Kapampangan, Bicol, Waray, Pangasinan, Spanish and Arabic (optional) Singapore English, Mandarin, Malay, Tamil, other Chinese languages, other Indian languages, Arabic dialects, others |
| South China Sea Islands | English, Filipino, Malay, Mandarin Chinese, Vietnamese |
| Thailand | Thai, English, Chinese languages, Malay, Lao, Khmer, Isaan, Shan, Lue, Phutai, Mon, Mein, Hmong, Karen, Burmese, others Vietnam Vietnamese, English, Chinese languages, French, Khmer, mountain area languages (Mon-Khmer and Malayo-Polynesian, hmong) |

### *Culture*

Rice paddy agriculture has existed in Southeast Asia for thousands of years, ranging across the subregion. Some dramatic examples of these rice paddies populate the Banaue Rice Terraces in the mountains of Luzon in the Philippines. Maintenance of these paddies is very labour-intensive. The rice paddies are well-suited to the monsoon climate of the region.

Stilt houses can be found all over Southeast Asia, from Thailand and Laos, to Borneo, to Luzon in the Philippines, to Papua New Guinea.

The region has diverse metalworking, especially in Indonesia. This include weaponry, such as the distinctive kris, and musical instruments, such as the gamelan.

### *Influences*

The region's chief cultural influences have been from either China or India or both, with Vietnam considered by far the most Chinese-influenced. Myanmar can be said to be influenced equally by both India and China. Western cultural influence is most pronounced in the Philippines, derived particularly from the period of Spanish rule.

As a rule, the peoples who ate with their fingers were more likely influenced by the culture of India, for example, than the culture of China, where the peoples first ate with chopsticks; tea, as a beverage, can be found across the region. The fish sauces distinctive to the region tend to vary.

### *The Arts*

The arts of Southeast Asia have no affinity with the arts of other areas. Dance in much of Southeast Asia also includes movement of the hands, as well as the feet to express the emotion and meaning of dance upon the story that the ballerina going to tell the audience. Most of Southeast Asian confirmed the Dance into their court, according to Cambodian royal ballet represent them in earlier of 7th century before Khmer Empire which highly influenced by Indian Hinduism. Apsara Dance, famous for its strongly hand and feet movement, is a great example of Hindism symbol dance. Puppetry and shadow plays were also a favoured form of entertainment in past centuries as the famous one known as Wayang from Indonesia. The Arts and Literature in some of Southeast Asia is quite influenced by Hinduism brought to them centuries ago.

The Tai, coming late into Southeast Asia, brought with them some Chinese artistic traditions, but they soon shed them in favour of the Khmer and Mon traditions, and the only indications of their earlier contact with Chinese arts were in the style of their temples, especially the tapering roof, and in their lacquerware.

In Indonesia, despite conversion to Islam opposed to certain forms of art, they retained many forms of Hindu influenced practices, cultures, arts and literatures. An example will be the Wayang Kulit (Shadow Puppet) and literatures like the Ramayana. This is also true for mainland Southeast Asia (excluding Vietnam). Dance movements, Hindu gods, arts were also fused into Thai, Khmer, Lao and Burmese cultures. It has been pointed out that Khmer and Indonesian classical arts were concerned with depicting the life of the gods, but to the Southeast Asian mind the life of the gods was the life of the peoples themselves—joyous, earthy, yet divine.

In Vietnam, the Vietnamese share many cultural similarities with the Chinese.

### *Music*

Traditional music in Southeast Asia is as varied as its many ethnic and cultural divisions. Main styles of traditional music can be seen: Court music, folk music, music styles of smaller ethnic groups, and music influenced by genres outside the geographic region.

Of the court and folk genres, Gong-chime ensembles and orchestras make up the majority (the exception being lowland areas of Vietnam). *Gamelan* orchestras from Indonesia, *Piphat/Pinpeat* ensembles of Thailand & Cambodia and the *Kulintang* ensembles of the southern Philippines, Borneo, Sulawesi and Timor are the three main distinct styles of musical genres that have influenced other traditional musical styles in the region. String instruments also are popular in the region.

### *Writing*

The history of Southeast Asia has led to a wealth of different authors, from both within and without writing about the region.

Originally, Indians were the ones who taught the native inhabitants about writing. This is shown through Brahmic forms of writing present in the region such as the Balinese script shown on split palm leaf called *lontar*, right:

The antiquity of this form of writing extends before the invention of paper around the year 100 in China. Note each palm leaf section was only several lines, written longitudinally across the leaf, and bound by twine to the other sections. The outer portion was decorated. The alphabets of Southeast Asia tended to be abugidas, until the arrival of the Europeans, who used words that also ended in consonants, not just vowels. Other forms of official documents, which did not use paper, included Javanese copperplate scrolls. This would have been more durable in the tropical climate of Southeast Asia.

# 3

# China

China is seen variously as an ancient civilization extending over a large area in East Asia, a nation and/or a multinational entity.

China is one of the world's oldest civilizations and is regarded as the oldest continuous civilization. From 1000 to 1400, it possessed one of the most advanced society and economy in the world through successive dynasties though it subsequently missed the industrial revolution and began to decline. In the 19th and 20th century, imperialism, internal weakness and civil wars damaged the country and its economy and led to the overthrow of imperial rule.

In 1949, when major combat ended in the Chinese Civil War, two political entities emerged having the term "China" in their names:

- The People's Republic of China (PRC), established in 1949, commonly known as *China*, has control over mainland China and the largely self-governing territories of Hong Kong (since 1997) and Macau (since 1999).
- The Republic of China (ROC) established in 1912 in mainland China, now commonly known as *Taiwan*, has control over the islands of Taiwan, Penghu, Kinmen, and Matsu.

In the 1950s, change to economic policies in Taiwan transformed the island into a technology-oriented industrialized developed economy after a period of high growth rates and rapid industrialization. In mainland China, in the 1970s, reforms known as the Four Modernizations modernized the agriculture, industry, technology and defence, vastly raising living standards, and making the PRC one of the great powers.

Historically, China's cultural sphere has extended across East Asia as a whole, with Chinese religion, customs, and writing systems being adopted to varying degrees by neighbours such as Japan, Korea and Vietnam. Through its history, China was the source of many major inventions. It has also one of the world's oldest written language systems. The first evidence of human presence in the region was found at the Zhoukoudian cave. It is one of the earliest known specimens of *Homo erectus*, now commonly known as the Peking Man, estimated to have lived from 300,000 to 780,000 years ago.

### *Etymology*

The traditional and simplified characters for "China" in Chinese. The first character means "middle" or "centre", and the second character means "country".

### *English Names*

The word "China" is derived from *Cin*, a Persian name for China popularized in Europe by Marco Polo. In early usage, "china" as a term for porcelain was spelled differently from the name of the country, the two words being derived from separate Persian words. Both these words are derived from the Sanskrit word *Cina*, used as a name for China as early as AD 150. The origin of this word is the subject of several conflicting scholarly theories. The traditional theory, proposed in the 17th century by Martin Martini, is that the word is derived from "Qin" (778 BC – 207 BC), the westernmost of the Chinese kingdoms during the Zhou dynasty, or from the succeeding Qin dynasty (221 – 206 BC). In the Hindu scriptures *Mahabharata* (5th century BC) and *Laws of Manu* (2nd century BC), the word *Cina* is used to refer to a country of "yellow-coloured" barbarians located in the Tibeto-Burman borderlands east of India.

## Chinese Names

The official name of China changed with each dynasty. The common name is *Zhongguo*. This translates traditionally as "Middle Kingdom," or as "central country." The name *Zhongguo* first appeared in the *Classic of History* (6th century BC), and was used to refer to the late Zhou Dynasty, as they believed that they were the "centre of civilization," while peoples in the four cardinals were called Eastern Yi, Southern Man, Western Rong and Northern Di respectively. Some texts imply that "Zhongguo" was originally meant to refer to the capital of the sovereign, to differ from the capital of his vassals. The use of *"Zhongguo"* implied a claim of political legitimacy, and *"Zhongguo"* was often used by states who saw themselves as the sole legitimate successor to previous Chinese dynasties; for example, in the era of the Southern Song Dynasty, both the Jin Dynasty and the Southern Song state claimed to be *"Zhongguo."*

*Zhongguo* came to official use as an abbreviation for the Republic of China (*Zhonghua Minguo*) after the government's establishment in 1912. Since the People's Republic of China, established in 1949, now controls the great majority of the area encompassed within the traditional concept of "China", the People's Republic is the political unit most commonly identified with the abbreviated name *Zhongguo*, with the Republic of China nowadays known commonly as "Taiwan".

## *History*

Ancient China was one of the earliest centers of human civilization. Chinese civilization was also one of the few to invent writing, the others being Mesopotamia, the Indus Valley civilization, the Maya and other Mesoamerican civilizations, the Minoan civilization of ancient Greece, and Ancient Egypt.

## *Prehistory*

Archaeological evidence suggests that the earliest hominids in China date from 250,000 to 2.24 million years ago. A cave in Zhoukoudian (near present-day Beijing) has fossils dated at somewhere between 300,000 to 780,000 years. The fossils are of Peking Man, an example of *Homo erectus* who used fire.

The earliest evidence of a fully modern human in China comes from Liujiang County, Guangxi, where a cranium has been found and dated at approximately 67,000 years old. Although much controversy persists over the dating of the Liujiang remains, a partial skeleton from Minatogawa in Okinawa, Japan has been dated to 16,600 to 18,250 years old, so modern humans probably reached China before that time.

## *Dynastic Rule*

Chinese tradition names the first dynasty Xia, but it was considered mythical until scientific excavations found early bronze-age sites at Erlitou in Henan Province in 1959. Archaeologists have since uncovered urban sites, bronze implements, and tombs in locations cited as Xia's in ancient historical texts, but it is impossible to verify that these remains are of the Xia without written records from the period.

The first Chinese dynasty that left historical records, the loosely feudal Shang (Yin), settled along the Yellow River in eastern China from the 17th to the 11th century BC. The Oracle bone script of the Shang Dynasty represent the oldest forms of Chinese writing found and the direct ancestor of modern Chinese characters used throughout East Asia. The Shang were invaded from the west by the Zhou, who ruled from the 12th to the 5th century BC, until their centralized authority was slowly eroded by feudal warlords. Many independent states eventually emerged out of the weakened Zhou state, and continually waged war with each other in the Spring and Autumn period, only occasionally deferring to the Zhou king. By the time of the Warring States period, there were seven powerful sovereign states, each with its own king, ministry and army.

The first unified Chinese state was established by Qin Shi Huang of the Qin state in 221 BC, who proclaimed himself as the "First Emperor" and created many reforms in the Empire, notably the forced standardization of the Chinese language and measurements. The Qin Dynasty lasted only fifteen years, as its harsh legalist and authoritarian policies soon led to widespread rebellion.

The subsequent Han Dynasty ruled China between 206 BC and 220 AD, and created a lasting Han cultural identity among its populace that extends to the present day. The Han Dynasty expanded the empire's territory considerably with military campaigns reaching Korea, Vietnam, Mongolia and Central Asia, and also helped establish the Silk Road in Central Asia.

After Han's collapse, another period of disunion followed, including the highly chivalric period of the Three Kingdoms. Independent Chinese states of this period such as Wu opened diplomatic relations with Japan, introducing the Chinese writing system there. In 580 AD, China was reunited under the Sui. However, the Sui Dynasty was short-lived after a failure in the Goguryeo-Sui Wars (598–614) weakened it.

Under the succeeding Tang and Song dynasties, Chinese technology and culture reached its zenith. The Tang Empire was at its height of power until the middle of the 8th century, when the An Shi Rebellion destroyed the prosperity of the empire. The Song dynasty was the first government in world history to issue paper money and the first Chinese polity to establish a permanent standing navy. Between the 10th and 11th centuries, the population of China doubled in size. This growth came about through expanded rice cultivation in central and southern China, and the production of abundant food surpluses.

Within its borders, the Northern Song Dynasty had a population of some 100 million people. The Song Dynasty was a culturally rich period for philosophy and the arts. Landscape art and portrait painting were brought to new levels of maturity and complexity after the Tang Dynasty, and social elites gathered to view art, share their own, and trade precious artworks. Philosophers such as Cheng Yi and Chu Hsi reinvigorated Confucianism with new commentary, infused Buddhist ideals, and emphasized a new organization of classic texts that brought about the core doctrine of Neo-Confucianism.

In 1271, the Mongol leader and fifth Khagan of the Mongol Empire Kublai Khan established the Yuan Dynasty, with the last remnant of the Song Dynasty falling to the Yuan in 1279. Before the Mongol invasion, Chinese dynasties reportedly had approximately 120 million inhabitants; after the conquest was completed in 1279, the 1300 census reported roughly 60 million people. A peasant named Zhu Yuanzhang overthrew the Mongols in 1368 and founded the Ming Dynasty. Ming Dynasty thinkers such as Wang Yangming would further critique and expand Neo-Confucianism with ideas of individualism and innate morality that would have tremendous impact on later Japanese thought. Chosun Korea also became a nominal vassal state of Ming China and adopted much of its Neo-Confucian bureaucratic structure.

Under the Ming Dynasty, China had another golden age, with one of the strongest navies in the world, a rich and prosperous economy and a flourishing of the arts and culture. It was during this period that Zheng He led explorations throughout the world, possibly reaching America. During the early Ming Dynasty China's capital was moved from Nanjing to Beijing. In 1644 Beijing was sacked by a coalition of rebel forces led by Li Zicheng, a minor Ming official turned leader of the peasant revolt. The last Ming Emperor Chongzhen committed suicide when the city fell. The Manchu Qing Dynasty then allied with Ming Dynasty general Wu Sangui and overthrew Li's short-lived Shun Dynasty, and subsequently seized control of Beijing, which became the new capital of the Qing dynasty.

The Qing Dynasty, which lasted until 1912, was the last dynasty in China. In the 19th century the Qing Dynasty adopted a defensive posture towards European imperialism, even though it engaged in imperialistic expansion into Central Asia. At this time China awoke to the significance of the rest of the world, the West in particular. As China opened up to foreign trade and missionary activity, opium produced by British India was forced onto Qing China. Two Opium Wars with Britain weakened the Emperor's control. European imperialism proved to be disastrous for China:

The weakening of the Qing regime, and the apparent humiliation of the unequal treaties in the eyes of the Chinese people had several consequences. One consequence was the Taiping Civil War, which lasted from 1851 to 1862. It was led by Hong Xiuquan, who was partly influenced by an idiosyncratic interpretation of Christianity. Hong believed himself to be the son of God and the younger brother of Jesus. Although the Qing forces were eventually victorious, the civil war was one of the bloodiest in human history, costing at least 20 million lives (more than the total number of fatalities in the First World War), with some estimates of up to two hundred million. Other costly rebellions followed the Taiping Rebellion, such as the Punti-Hakka Clan Wars (1855–67), Nien Rebellion (1851–1868), Muslim Rebellion (1862–77), Panthay Rebellion (1856–1873) and the Miao Rebellion (1854–73).

These rebellions resulted in an estimated loss of several million lives each and led to disastrous results for the economy and the countryside. The flow of British opium hastened the empire's decline. In the 19th century, the age of colonialism was at its height and the great Chinese Diaspora began. About 35 million overseas Chinese live in Southeast Asia today. The famine in 1876–79 claimed between 9 and 13 million lives in northern China. From 108 BC to 1911 AD, China experienced 1,828 famines, or one per year, somewhere in the empire.

While China was wracked by continuous war, Meiji Japan succeeded in rapidly modernizing its military and set its sights on Korea and Manchuria. At the request of the Korean emperor, the Chinese government sent troops to aid in suppressing the Tonghak Rebellion in 1894. However, Japan also sent troops to Korea, leading to the First Sino-Japanese War, which resulted in Qing China's loss of influence in the Korean Peninsula as well as the cession of Taiwan to Japan.

Following this series of defeats, a reform plan for the empire to become a modern Meiji-style constitutional monarchy was drafted by the Emperor Guangxu in 1898, but was opposed and stopped by the Empress Dowager Cixi, who placed Emperor Guangxu under house arrest in a coup d'etat. Further destruction followed the ill-fated 1900 Boxer Rebellion against westerners in Beijing.

By the early 20th century, mass civil disorder had begun, and calls for reform and revolution were heard across the country. The 38-year-old Emperor Guangxu died under house arrest on 14 November 1908, suspiciously just a day before Cixi's own death. With the throne empty, he was succeeded by Cixi's handpicked heir, his two year old nephew Puyi, who became the Xuantong Emperor. Guangxu's consort, who became the Empress Dowager Longyu. In another coup d'etat, Yuan Shikai overthrew the last Qing emperor, and forced empress Dowager Longyu to sign the abdication decree as regent in 1912, ending two thousand years of imperial rule in China. She died, childless, in 1913.

### *Republic of China (1912–49)*

On 1 January 1912, the Republic of China was established, heralding the end of the Qing Dynasty. Sun Yat-sen of the

Kuomintang (the KMT or Nationalist Party) was proclaimed provisional president of the republic. However, the presidency was later given to Yuan Shikai, a former Qing general, who had ensured the defection of the entire Beiyang Army from the Qing Empire to the revolution. In 1915, Yuan proclaimed himself Emperor of China but was forced to abdicate and return the state to a republic when he realized it was an unpopular move, not only with the population but also with his own Beiyang Army and its commanders.

After Yuan Shikai's death in 1916, China was politically fragmented, with an internationally recognized but virtually powerless national government seated in Peking (Beijing). Warlords in various regions exercised actual control over their respective territories. In the late 1920s, the Kuomintang, under Chiang Kaishek, was able to reunity the country under its own control, moving the nation's capital to Nanking (Nanjing) and implementing "political tutelage", an intermediate stage of political development outlined in Sun Yatsen's program for transforming China into a modern, democratic state. Effectively, political tutelage meant one-party rule by the Kuomintang.

The Sino-Japanese War of 1937–1945 (part of World War II) forced an uneasy alliance between the Nationalists and the Communists as well as causing around 20 million Chinese civilian deaths. With the surrender of Japan in 1945, China emerged victorious but financially drained. The continued distrust between the Nationalists and the Communists led to the resumption of the Chinese Civil War. In 1947, constitutional rule was established, but because of the ongoing Civil War many provisions of the ROC constitution were never implemented in mainland China.

### *Post Civil War (1949–present)*

After its victory in the Chinese Civil War, the Communist Party of China (CCP) led by Mao Zedong gained control of most of Mainland China. On 1 October 1949, they established the People's Republic of China as a Socialist State headed by a "Democratic Dictatorship" with the CCP as the only legal political party, thus, laying claim as the successor state of the ROC. The central government of the Chinese Nationalist Party led by Chiang Kai-shek retreated to the island of Taiwan that it had occupied at the end of World War II, and moved the ROC government there. Major armed hostilities ceased in 1950 but no peace treaty has been signed. An estimated 36 million died during the Great Chinese Famine of 1958–61.

Beginning in the late 1970s, the Republic of China began the implementation of full, multi-party, representative democracy in the territories still under its control (Taiwan, and a number of smaller islands including Quemoy and Matsu). Today, the ROC has active political participation by all sectors of society. The main cleavage in ROC politics is the issue of eventual political unification with the Chinese mainland vs formal independence of Taiwan.

After the Chinese Civil War, mainland China underwent a series of disruptive socioeconomic movements starting in the late 1950s with the Great Leap Forward and continuing in the 1960s with the Cultural Revolution that left much of its education system and economy in shambles. With the death of its first generation Communist Party leaders such as Mao Zedong and Zhou Enlai, the PRC began implementing a series of political and economic reforms advocated by Deng Xiaoping that eventually formed the foundation for mainland China's rapid economic development starting in the 1990s.

Post-1978 reforms in mainland China have led to some relaxation of control over many areas of society. However, the PRC government still has almost absolute control over politics, and it continually seeks to eradicate what it perceives as threats to the social, political and economic stability of the country. Examples include the fight against terrorism, jailing of political opponents and journalists, custody regulation of the press, regulation of religion, and suppression of independence/ secessionist movements. In 1989, the student protests at Tiananmen Square were violently put to an end by the Chinese military after 15 days of martial law. In 1997, Hong Kong was ceded to the PRC by the United Kingdom, and in 1999, Macau was handed over by Portugal.

Since 1949, mainland China is administered by the People's Republic of China—a one-party state under the leadership of the Chinese Communist Party—while the island of Taiwan and surrounding islands are administered by the Republic of China—a democratic multi-party state. After the founding of the People's

Republic in 1949, both states claimed to be the sole legitimate ruler of all of China. After the Kuomintang retreat to Taiwan in 1949, the Republic of China had maintained official diplomatic relations with most states around the world, but by the 1970s, a shift had occurred in international diplomatic circles and the People's Republic of China gained the upper hand in international diplomatic relations and recognition count.

In 1971, under UN resolution 2758, the representatives of Chiang Kai-shek to the United Nations were expelled from the intergovernmental organization. With the expulsion of the representatives, and effectively the Republic of China, the representatives of the People's Republic of China were invited to assume China's seat on the UN Security Council, the UN General Assembly and other United Nations councils and agencies. Later attempts by the Republic of China to rejoin the UN have either been blocked by the People's Republic of China, which has veto power on the UN Security Council, or rejected by the United Nations Secretariat or a United Nations General Assembly committee responsible for the General Assembly's agenda.

Since the relocation of its capital to Taiwan, the Republic of China has not formally renounced its claim to authority over all of China, nor has it changed its official maps, which include the mainland and Mongolia. Following the introduction of full democracy, and the electoral victory of the DPP's Chen Shuibian in the presidential elections, the ROC had adopted a policy of separating the state's identity from "China", while moving towards identifying the state as "Taiwan".

However, the ROC has not made any formal moves to change the name, flag, or national anthem of the state to reflect a Taiwanese identity due to the lack of consensus within Taiwan, pressure from the United States and the fear of invasion or military action from the People's Republic of China against the island. The Republic of China during the DPP years did not actively pursue its claims on mainland China or Mongolia. However, after having been elected as president, KMT's Ma Yingjeou asserted that, constitutionally, mainland China is part of the Republic of China. The People's Republic of China claims to have succeeded the Republic of China as the sole legitimate governing authority of all of China, which, from the official viewpoint of the People's Republic of China, includes the island of Taiwan.

Over the last 50 years, both the Republic of China and the People's Republic of China have used diplomatic and economic means to compete for recognition in the international arena. Because most international, intergovernmental organizations observe the One-China policy of the People's Republic of China, the PRC has been able to pressure organizations, such as the World Health Organization and the International Olympic Committee, to refuse to officially recognize the Republic of China. Due to the One-China policy, states around the world are pressured to refuse, or to cut off diplomatic relations with the Republic of China. As a result, 23 U.N. member states currently maintain official diplomatic relations with the Republic of China, while the vast majority of U.N. member states maintain official diplomatic relations with the People's Republic of China.

## Territory and Environment

### *Historical Political Divisions*

Top-level political divisions of China have altered as administrations changed. Top levels included circuits and provinces. Below that, there have been prefectures, sub-prefectures, departments, commanderies, districts, and counties. Recent divisions also include prefecture-level cities, county-level cities, towns and townships.

Most Chinese dynasties were based in the historical heartlands of China, known as China proper. Various dynasties also expanded into peripheral territories like Inner Mongolia, Manchuria, Xinjiang, and Tibet. The Manchu-established Qing Dynasty and its successors, the ROC and the PRC, incorporated these territories into the Chinese empire.

## Geography of China

China stretches some 5,026 kilometres (3,123 mi) across the East Asian landmass bordering the East China Sea, Korea Bay, Yellow Sea, and South China Sea, between North Korea and Vietnam in a changing configuration of broad plains, expansive deserts, and lofty mountain ranges, including vast areas of inhospitable terrain. The eastern half of the country, its seacoast fringed with offshore islands, is a region of fertile lowlands, foothills and mountains, deserts, steppes, and subtropical areas. The western half of China is a region of sunken basins, rolling plateaus, and towering massifs, including a portion of the highest tableland on earth.

The vastness of the country and the barrenness of the western hinterland have important implications for defence strategy. In spite of many good harbours along the approximately 18,000-kilometer coastline, the nation has traditionally oriented itself not toward the sea but inland, developing as an imperial power whose centre lay in the middle and lower reaches of the Yellow River on the northern plains. China also has the Tibetan Plateau, a very large, high altitude plateau, to the south. To the north of the Tibetan Plateau lie the Gobi and Taklamakan deserts, which stretch from the extreme northwest eastward through Mongolia.

The People's Republic of China is one of the world's largest countries in total area behind Russia and Canada, and very similar to the United States. Whether China or the United States is the third largest country in the world in total area is related to (a) the validity of claims by the PRC on territories such as Taiwan, Aksai Chin, Trans-Karakoram Tract, and South Tibet (Aksai Chin and Trans-Karakoram Tract also claimed by India), and (b) how the total size of the United States is calculated: The CIA's *The World Factbook* gives 9,826,630 km$^2$, the United Nations Statistics Division gives 9,629,091 km$^2$, and the *Encyclopedia Britannica* gives 9,522,055 km$^2$. Figures for the size of China differ slightly depending on where one draws a number of ill-defined boundaries.

The official figure by the People's Republic of China is 9.6 million square kilometers. The Republic of China based in Taiwan but claiming to be the government of China puts this figure at 11 million square kilometers, but this includes Mongolia a state whose sovereignty has been recognized by the PRC. China's contour is reasonably comparable to that of the United States and lies largely at the same latitudes. The total area is estimated to be 9,758,801 km$^2$, with land accounting for 9,326,410 km$^2$ and water for 270,550 km$^2$ (around 3 percent).

## *Geography*

From the Tibetan Plateau and other less-elevated highlands rise rugged east-west trending mountains, and plateaus interrupted by deep depressions fanning out to the north and east. The Tibetan Plateau is a vast, elevated plateau covering most of the Tibet Autonomous Region and Qinghai Province in the People's Republic of China and Ladakh in India. With an average elevation of over 4,500 meters, the highest and biggest plateau in the world and an area of 2.5 million square kilometers. A continental scarp marks the eastern margin of this territory, a scarp that extends from the Greater Khingan Range in northeastern China, through the Taihang Mountains (a range of mountains overlooking the North China Plain) to the eastern edge of the Yunnan-Guizhou Plateau in the south. All of the low-lying areas of China, which support dense population and intensive cultivation, are to the east of this scarp line.

The east-west ranges include some of Asia's greatest mountains. In addition to the Himalayas and the Kunlun Mountains, there are the Mount Kailash (Gangdise) and the Tian Shan ranges. The latter stands between two great basins, the massive Tarim Basin to the south and the Dzungarian Basin to the north. Rich deposits of coal, oil, and metallic ores lie in the Tian Shan area. The largest inland basin in China, the Tarim Basin measures 1,500 kilometres from east to west and 600 kilometres from north to south at its widest parts. The Himalayas form a natural boundary on the southwest as the Altai Mountains do on the northwest. Lesser ranges branch out, some at sharp angles from the major ranges. The mountains give rise to all the principal rivers. The spine of the Kunlun Mountains separates into several branches as it runs eastward from the Pamir Mountains. The northernmost branches, the Altyn-Tagh and the Qilian Range, form the rim of the Tibetan Plateau in west-central China and overlook the Qaidam Basin, a sandy and swampy region containing many salt lakes. A southern branch of the Kunlun Mountains divides the watersheds of the Yellow River (Huang He) and the Yangtze River (Chang Jiang). The Gansu Corridor, west of the great bend in the Yellow River, was traditionally an important communications link with Central Asia.

North of the 3,300-kilometre-long Great Wall, between Gansu Province on the west and the Greater Khingan Range on the east, lies the Mongolian Plateau, at an average elevation of 1,000 metres above sea level. The Yin Mountains, a system of mountains with average elevations of 1,400 metres, extends east-west through the centre of this vast desert steppe. To the south is the largest loess plateau in the world, covering 600,000 square kilometers in Shaanxi Province, parts of Gansu and Shanxi provinces, and some of Ningxia-Hui Autonomous Region. Loess is a yellowish soil blown in from the Inner Mongolian deserts. The loose, loamy material travels easily in the wind, and through the centuries it has veneered the plateau and choked the Yellow River with silt. Because the river level drops precipitously toward the North China Plain where it sluggishly crosses the delta, it carries a heavy load of sediment in the form of sand and mud from the upper reaches, much of which is deposited on the flat plain. The flow is controlled mainly by constantly repaired man-made embankments while floods and course changes have recurred over the centuries. As a result the river flows on a raised ridge fifty metres or more above the plain, Traditionally, rulers were judged by their concern for or indifference to preservation of the embankments.

Flowing from its source in the Tibetan highlands, the Yellow River courses toward the sea through the North China Plain, the historic centre of Chinese expansion and influence. Ethnic Chinese people have farmed the rich alluvial soils of the plain since ancient times, constructing the Grand Canal of China for north-south transport. The plain itself is actually a continuation of the Manchurian Plain to the northeast but is separated from it by the Bohai Gulf, an extension of the Yellow Sea. Like other densely populated areas of China, the plain is subject not only to floods but to earthquakes. For example, the mining and industrial centre of Tangshan, about 165 kilometres east of Beijing, was levelled by an earthquake in July 1976 that reportedly also killed 242,000 people and injured 164,000.

The Qinling mountain range, a continuation of the Kunlun Mountains, divides the North China Plain from the Yangtze River Delta and is the major physiographic boundary between the two great parts of China Proper. It is in a sense a cultural boundary as well, influencing the distribution of custom and language. South of the Qinling divide are the densely populated and highly developed areas of the lower and middle plains of the Yangtze and, on its upper reaches, the Sichuan Basin, an area encircled by a high barrier of mountain ranges. The country's longest and most important waterway, the Yangtze River is navigable over much of its length and is now the site of the Three Gorges Dam. Rising on the Tibetan Plateau, the Yangtze River traverses 6,300 kilometers through the heart of the country, draining an area of 1.8 million square kilometers before emptying into the East China Sea. The Sichuan Basin, favoured by a mild, humid climate and a long growing season, produces a rich variety of crops; it is also a leading silk-producing area and an important industrial region with substantial mineral resources.

Second only to the Qinling as an internal boundary is the Nanling, the southernmost of the east-west mountain ranges. The Nanling overlooks the part of China where a tropical climate permits two crops of rice to be grown each year. Southeast of the mountains lies a coastal, hilly region of small deltas and narrow valley plains; the drainage area of the Pearl River (Zhu Jiang) and its associated network of rivers occupies much of the region to the south. West of the Nanling, the Yunnan-Guizhou Plateau rises in two steps, averaging 1,200 and 1,800 metres in elevation, respectively, toward the precipitous mountain regions of the eastern Tibetan Plateau.

The Hai River, like the Pearl and other major waterways, flows from west to east. Its upper course consists of five rivers that converge near Tianjin, then flow seventy kilometers before emptying into the Bohai Gulf. Another major river, the Huai, rises in Henan Province and flows through several lakes before joining the Yangtze near Yangzhou. Inland drainage involving a number of upland basins in the north and northeast accounts for about 40 percent of the country's total drainage area. Many rivers and streams flow into lakes or diminish in the desert. Some are useful for irrigation.

China's extensive territorial waters are principally marginal seas of the western Pacific Ocean; these waters wash the shores of a long and much-indented coastline and approximately 5,000 islands. The Yellow, East China, and South China seas, too, are marginal seas of the Pacific Ocean. More than half the coastline (predominantly in the south) is rocky; most of the remainder is sandy. Hangzhou Bay roughly divides the two kinds of shoreline.

Areas of China have experienced earthquakes. On 23 August 1976, a major earthquake in Tangshan killed hundreds of thousands of people. However, most regions of China do not experience earthquakes, as major population centres are a long distance from fault lines. Tangshan is one of the few places in China that is located within an earthquake zone. There are few volcanoes in China.

### *Topography*

The topography varies greatly in China, a vast land of lofty plateau, large plains, rolling land and big and small basins surrounded by lofty mountains. All the five basic topographic types in the world exist in China to create the conditions for developing industry and agriculture.

Mountainous land and very rough terrains make up two-thirds of Chinese territory, and this has created some problems in transport and in the development of agricultural production. However such topographical features are conducive to the development of forestry, mineral and hydropower resources and tourism.

With highlands in the west and plains in the east, China has a varied topography. The plane of the land may be divided into three tiers. The Qinghai-Tibet Plateau that rises more than 4000m above sea-level forms the highest tier. It is a land of peaks and valleys studded with innumerable lakes. Along the plateau's southwestern fringe is the Himalayan Range, on the eastern section of which looms the 8,848.13 meter-high Mt. Everest, the world's highest peak.

The vast area north and east of the Qinghai-Tibet Plateau that drops to an elevation below 1,100-2,000m forms the second tier-a land interspersed with extensive basins and highlands. Here the Turpan Basin in Xinjiang is 154m below sea-level-the lowest depression in China. The third tier is a vast area of rolling hills and plains with an elevation below 500m lying east of the line running from the Greater Hinggan and Taihang ranges in the north to the foothills of the Wushan Mountains and the Yunnan-Guizhou Highlands in the south. Though some peaks in this area are as high as 2,000m, the plains along the coast have an elevation of less than 50m. Off the Chinese coast is an extensive continental shelf richly endowed with petroleum, natural gas and marine resources.

There are many mountain ranges in China. Those extending from east to west are the Tian Shan-Yinshan ranges and those in the centre are the Kunlun-Qinling ranges, and those in the south are the Nanling ranges. Ranges that stretch in a northeast-southwest direction are, for the most part, located in the eastern part of China. They are the Greater Hinggan Range, Taihang, Wushan, Xuefengshan, Changbaishan and Wuyishan ranges. Those running in a NW-SE direction are the Altai mountains and Qilianshan ranges. Ranges that run in a north-south direction are the Hengduanshan and Helanshan ranges. And on the border between China and India, Nepal and other countries looms the 2400 kilometers-long Himalayan Range with an average elevation of 6,000m.

### *Climate*

The climate of China is extremely diverse; subtropical in the south to subarctic in the north. Monsoon winds, caused by differences in the heat-absorbing capacity of the continent and the ocean, dominate the climate. Alternating seasonal air-mass movements and accompanying winds are moist in summer and dry in winter. The advance and retreat of the monsoons account in large degree for the timing of the rainy season and the amount of rainfall throughout the country. Tremendous differences in latitude, longitude, and altitude give rise to sharp variations in precipitation and temperature within China. Although most of the country lies in the temperate belt, its climatic patterns are complex.

China's northernmost province Heilongjiang has a subarctic climate; its southernmost point, Hainan Island (an island away from mainland China), has a tropical climate. Temperature differences in winter are great, but in summer the diversity is considerably less. For example, the northern portions of Heilongjiang Province experience an average January mean temperature of below 0 °C (32 °F), and the reading may drop to "30 °C ("22.0 °F); the average July mean in the same area may exceed 20 °C (68 °F).

By contrast, the central and southern parts of Guangdong Province experience an average January temperature of above 10 °C (50 °F), while the July mean is about 28 °C (82 °F).

Precipitation varies regionally even more than temperature. China south of the Qinling mountains experiences abundant rainfall, most of it coming with the summer monsoons. To the north and west of the range, however, rainfall is uncertain. The farther north and west one moves, the scantier and more uncertain it becomes. The northwest has the lowest annual rainfall in the country and no precipitation at all in its desert areas. China experiences frequent typhoons (about five per year along southern and eastern coasts), damaging floods, monsoons, tsunamis, and droughts.

## Hot Summer Cold Winter Zone

Hot-summer/cold-winter zone is the transient climate region between the cold and the hot zones in China.

It includes the whole of Hubei, Hunan, Jiangxi, Anhui, Zhejiang provinces, Shanghai and Chongqing two municipalities, the eastern part of Sichuan and Guizhou provinces, the southern part of Henan, Jiangsu, Shanxi and Gansu provinces, and the northern part of Fujian, Guangdong and Guangxi provinces. The zone includes an area of 1,800,000 km$^2$ (694,984 sq mi) with a population of 550 million people. This region is the most populous and economical-developed area of China, producing 48% of the gross domestic product (GDP) of the whole country.

The main bother of hot-summer/cold-winter zone is hot, humid summers and cold, humid winters. The temperature difference between day and night is normally small. The precipitation in an average year is large. Sun radiation is relative weak due to cloud cover.

The most durable time you will ever have outside during the hottest summer month is 25–30 °C (77–86 °F), with peak temperatures above 40 °C (104 °F). The average outside temperature during the coldest winter month is 0–10 °C (32–50 °F), with lowest temperatures below 0 °C (32 °F).

For historical reasons, the residential buildings in this zone don't have central HVAC systems and are not well insulated or otherwise weatherized. With the recent and rapid economic development of this region, demand for better indoor environments is rising. Many residents install 'minisplit' air conditioners to improve their thermal comfort. But electrical energy consumption is rising accordingly and is taxing the generation capacity. The Chinese government has created new national design standards and other efforts to lower the energy consumption while also constructing new power generating stations.

## Environment

The scale of China's environmental problems is large. To begin with, China's 1.3 billion human population account for around a fifth of the world's population, but the nation encompasses less than one tenth of the world's arable land. Furthermore, almost the entire population lives in the well-watered eastern half of the country, where virtually every square centimeter of farmland has been developed. Indeed, China has very little land that has not been altered in some way by man.

The sheer size of the population means that forests, wetlands, grasslands and agricultural fields are stretched beyond the limits of sustainable use. Dramatic growth in the economy and the continuing need to raise living standards for some of Asia's poorest people means that urban areas face a similar crisis: coal dust, untreated factory emissions, vehicle exhaust and wind-blown desert sand make Chinese cities some of the most polluted on Earth; many of the nation's rivers are polluted and virtually all water in urban areas is heavily contaminated.

Air pollution (sulfur dioxide particulates) from reliance on coal is a major issue, along with water pollution from untreated wastes and use of debated standards of pollutant concentration rather than Total Maximum Daily Load. There are water shortages, particularly in the north. The eastern part of China often experiences smoke and dense fog in the atmosphere as a result of industrial pollution. Heavy deforestation with an estimated loss of one-fifth of agricultural land since 1949 to soil erosion and economic development is occurring with resulting desertification. The size of the Gobi desert has increased and now reaches the outskirts of Beijing.

China is a party to the Antarctic-Environmental Protocol, the Antarctic Treaty, the Convention on Biological Diversity, the Climate Change treaty, the United Nations Convention to Combat Desertification, the Endangered Species treaty, the Hazardous Wastes treaty, the Law of the Sea, the International Tropical Timber Agreements of 1983 and 1994, the International Convention for the Regulation of Whaling, and agreements on Marine Dumping, Ozone Layer Protection, Ship Pollution, and Wetlands protection. China has signed, but not ratified the Kyoto Protocol (but is not yet required to reduce its carbon emission under the agreement, as is India), and the Nuclear Test Ban treaty.

### *Habitats*

Arguably the world's third largest country, China rises from sea level in the east to peak of Mount Everest on the border with Nepal. The south shares tropical rain forests with Laos, Vietnam and Burma, while the Da Hinggan Mountains in Inner Mongolia have tundra vegetation on top of permafrost. China is also home to East Asia's most important wetlands and Asia's longest river (Yangtze), and is the source of two rivers of inestimable importance to hundreds of millions of people in South and Southeast Asia-the Indus and the Mekong. Deserts make up one-fifth of China's total territory, largely in the northwest. Arid steppes cover additional areas in the Altai, Tian, and Kunlun Mountains in the far west, a region blocked from the southwestern monsoon by the Tibetan plateau and from the southeastern monsoon by its distance from the sea. This massive diversity of geography and habitats has resulted in an extraordinary range of plant and animal life.

### *Forests*

China contains a variety of forest types. Both northeast and northwest reaches contain mountains and cold coniferous forests, supporting animal species which include moose and Asiatic black bear, along with some 120 types of birds. Moist conifer forests can have thickets of bamboo as an understorey, replaced by rhododendrons in higher montane stands of juniper and yew. Subtropical forests, which dominate central and southern China, support an astounding 146,000 species of flora, as well as the famous giant panda, golden monkey and South

China tiger. Tropical rainforest and seasonal rainforests, though confined to Yunnan and Hainan Island, actually contain a quarter of all the plant and animal species found in China.

### *Grasslands*

Grasslands make up about a third of China's total land area. The immense and productive grasslands are largely concentrated in inner Mongolia, Ningxia, parts of Qinghai and Tibet. The natural wildlife they support includes three species on the verge of extinction: Przewalski's horse, the Asiatic wild ass and the Bactrian camel (the ancestor of domesticated camels). There is often direct competition between domestic animals and wild fauna, and herdsmen poison or trap carnivores, and sometimes set fires to increase pasture area. The government has recently stepped up efforts to control the conversion of grasslands to pasture, but lacks the manpower to enforce policy. As the area of grassland in China began to decrease, which inturn cause the closure of the biological home of several endangered species. Qinghai and Ningxia, are the home of several animals that already extincted.

### *Freshwater Ecosystems*

Freshwater habitats are of massive importance to China, and a huge percentage of the population is directly dependent on wetlands — marshes, rivers, and lakes — for economic activity, flood control and drinking water. Seven of the most important rivers in the world begin in the highlands of western China. The Yellow River (Huang He), Yangtze River (Chang Jiang), Lancang Jiang (Mekong) and the Salween rise in the east of the Qinghai-Tibet Plateau. The Indus and Brahmaputra rise in the south. Downstream these rivers serve as sources of irrigation and drinking water, modes of transportation, and centers of cultural and religious importance for some two billion people in China, India, Pakistan, Bangladesh and throughout Southeast Asia. These rivers rise and gather strength from many of the thousands of freshwater lakes of the region.

China's northeast is the focus for much of the country's freshwater marshes. An area of 20,000 square kilometres on the Sanjiang Plain of Heilongjiang is essentially a collection of shallow freshwater lakes and reed-beds where the Heilongjiang, Sungari, and Wusuli rivers come together. Jilin, Liaoning and

Inner Mongolia all share these ecosystems. One of the most well-known wildlife areas in this ecosystem is Zhalong Nature Reserve, a 2,000-square-kilometer area which was created in 1979 to protect breeding areas for the red-crowned crane, and other wintering migrants. These marshes are also of great value for reed production, the bulk of which is turned into pulp for paper. Waterfowl and reed production can usually coexist, at least at present levels, so this is a useful confluence of conservation and economic uses. In western Sichuan, marshland provides breeding grounds for the black-necked crane and bar-headed goose.

China's freshwater lakes include the country's best-known wetlands: Jiangxi's Poyang Hu and Hunan's Dongting Hu. Dongting Hu, China's second largest freshwater lake, is vitally important for wildlife, including the likely extinct Yangtze river dolphin and the highly endangered Chinese sturgeon, as well as more wintering wildfowl. Poyang Hu is a similar complex of small lakes and marsh areas which fluctuates seasonally; summer floods give way in autumn to fertile agricultural land, attractive both to farmers and visiting birds. The importance of the area is hard to overstate, as the lakes provide a wintering habitat for almost the entire world population of two hundred Siberian Cranes, and as many as five hundred thousand birds may be on Poyang Hu at any one time during the winter months. In recent years, however, some of Poyang's larger lakes have been drained at the end of autumn, leaving waterfowl with inadequate shallow land on which to feed.

### *In History*

Chinese history is often explained in terms of several strategic areas, definite by particular topographic limits. Starting from the Chinese central plain, the former heart of the Han populations, the Han people expanded militarily and then demographically toward the Loess plateau, the Sichuan plain, and the Southern hills, not without resistance from local populations. Pushed by its comparatively higher demographic growth, the Han continued their expansion by military and demographic waves. The far-south of nowadays China, the northern parts of today's Vietnam, the Tarim basin were first reached and durably subdued by Han's dynasty's armies. The northern steppes (Mongolia, Siberia), and the Tibetan plateau were softly vassalized in Tang times, and strongly integrated under the Yuan dynasty. The Manchurian plain and Korean peninsula were usually not under Chinese contol, with the exception of some limited periods of occupation. The Manchurian plain became strongly integrated into the Chinese empire during the late Qing dynasty, while the West side of the Changbai mountains, formerly the home of Korean tribes, thus also entered China.

Later 19th and 20th centuries struggles led the Chinese Empire to collapse under westerners pressures, and the PRC to rise under its today shape.

### *Saltwater Lakes*

About half of China's lakes are saline and, once again, are important breeding grounds for waterfowl. Most are concentrated in northwest China on the inland drainage systems of the North Tibetan Plain and in the Zaidan basin. The largest is Qinghai Lake, a 4,426-square-kilometer reserve which attracts thousands of birds each summer, including cormorants, great black-headed gulls, bar-headed geese and pied avocets. Similarly, the Tarim River basin in Xinjiang supports one of the largest breeding populations of black stork in China. The Ordos plateau area of Inner Mongolia as well as the Xinjiang's Taolimiao-Alashan Nur (lake) support breeding sites for the endangered relict gull. Most of these lakes and marshes fluctuate seasonally and are threatened by increased diversion of water for human use.

### *Coastal Wetlands*

China's coastline is approximately 18,000 km long, extending from the Bohai Gulf, which freezes in the winter, to the tropical waters of the South China Sea. Coastal wetlands are important as fuel stops for waterfowl on the migratory route between Siberia and Australia. Chongming Island in the Yangtze River delta near Shanghai-China's largest city and one of its fastest growing regions-is vital for these migrants.

### *Wildlife*

China lies in two of the world's major ecozones, the Palearctic and the Indomalaya. In the Palearctic zone are found such important mammals as the horse, camel, tapir, and jerboa.

Among the species found in the Indomalaya region are the Leopard Cat, bamboo rat, treeshrew, and various other species of monkeys and apes. Some overlap exists between the two regions because of natural dispersal and migration, and deer or antelope, bears, wolves, pigs, and rodents are found in all of the diverse climatic and geological environments. The famous giant panda is found only in a limited area along the Chang Jiang. There is a continuing problem with trade in endangered species, although there are now laws to prohibit such activities.

## Economic History of Modern China

The economic history of modern China began with the fall of the Qing Dynasty in 1911. Following the Qing, China underwent a period of instability and disrupted economic activity. Under the Nanjing decade (1927–1937), China advanced several industries, in particular those related to the military, in an effort to catch up with the west and prepare for war with Japan. The Second Sino-Japanese War (1937–1945) and the following Chinese civil war caused the collapse of the Republic of China and formation of the People's Republic of China.

The new ruler of China, Mao Zedong, initially promised to develop "an socialist alliance with petit-bourgeois, workers, and nationalist bourgeois", but enacted collectivization upon consolidation of this regime. Collectivization resulted in the success of the first five-year plan, but Mao's second five-year plan, which included the Great Leap forward, did not meet with the same success. A new party faction who supported private plots eventually challenged Mao's economic policy. Unwilling to give up power, Mao launched the Cultural Revolution, which led to the collapse of the Chinese economy.

Following Mao's death, one of the most senior officials who had advocated private plots in the early 1960s, Deng Xiaoping, initiated gradual market reforms that abolished the communes and collectivized industries of Mao, replacing them with the free-market system. Deng's reforms vastly improved the standard of living of the Chinese people, the competitiveness of the Chinese economy, and caused China to become one of the fastest growing and most important economies in the world. It also led to one of the most rapid industrializations in world history. For this achievement he is sometimes known as "The

Venerated Deng". As a result of Deng's reforms, China is widely regarded as a returning superpower.

### *Republic of China (1911-1949)*

The Republic of China was a period of turmoil for China after the collapse of the Qing dynasty. From 1911 to 1927, China virtually disintegrated into regional warlords, fighting for authority and causing economic misery and contraction. After 1927, Chiang Kai-shek managed to reunity China and bring in the Nanjing decade, a period of relative prosperity despite civil war and Japanese aggression. In 1937, the Japanese invaded and literally laid China to waste in eight years of war. The era also saw the first boycott of Japanese products. Afterwards, the Chinese civil war further devastated China and led to the fall of the Republic in 1949.

### *Civil War, Famine and Turmoil in the Early Republic*

The early republic was marked by frequent wars and factional struggles. From 1911 to 1927, famine, war and change of government was the norm in Chinese politics, with provinces periodically declaring "independence". The collapse of central authority caused the economic contraction that was in place since Qing to speed up, and was only reversed when Chiang reunified China in 1927 and proclaimed himself its leader.

### *Development of Domestic Industries*

Chinese domestic industries developed rapidly after the downfall of the Manchu Qing dynasty, despite turmoil in Chinese politics. Development of these industries peaked during World War I, which saw a great increase in demand for Chinese goods, which benefitted China's industries. In addition, imports to China fell drastically after total war broke out in Europe. For example, China's textile industry had 482,192 needle machines in 1913, while by 1918 (the end of the war) that number had gone up to 647,570. The number increased even faster to 1,248,282 by 1921. In addition, bread factories went up from 57 to 131.

The May 4th movement, in which Chinese students called China's population to boycott foreign goods, also helped spur development. Foreign imports fell drastically from 1919–1921 and from 1925 to 1927.

Chinese industries continue to develop in the 1930s with the advent of the Nanking decade in the 1930s, when Chiang Kai-shek unified most of the country and brought political stability. China's industries developed and grew from 1927 to 1931. Though badly hit by the Great Depression from 1931 to 1935 and Japan's occupation of Manchuria in 1931, industrial output recovered by 1936. By 1936, industrial output had recovered and surpassed its previous peak in 1931 prior to the Great Depression's effects on China. This is best shown by the trends in Chinese GDP. In 1932, China's GDP peaked at 28.8 billion, before falling to 21.3 billion by 1934 and recovering to 23.7 billion by 1935.

### *The Rural Economy of the Republic of China*

The rural economy retained much of the characteristics of the Late Qing. While markets had been forming since the Song and Ming dynasties, Chinese agriculture by the Republic of China was almost completely geared towards producing cash crops for foreign consumption, and was thus subject to the say of the international markets. Key exports included glue, tea, silk, sugar canes, tobacco, cotton, corn and peanuts.

The rural economy was hit hard by the Great Depression of the 1930s, in which an overproduction of agricultural goods lead to massive falling prices for China as well as an increase in foreign imports (as agricultural goods produced in western countries were "dumped" in China). In 1931, imports of rice in China amounted to 21 million bushels compared with 12 million in 1928. Other goods saw even more staggering increases. In 1932, 15 million bushels of grain were imported compared with 900,000 in 1928. This increased competition lead to a massive decline in Chinese agricultural prices (which were cheaper) and thus the income of rural farmers. In 1932, agricultural prices were 41 percent of 1921 levels.. Rural incomes had fallen to 57 percent of 1931 levels by 1934 in some areas.

### *Foreign Direct Investment in the Republic of China*

Foreign direct investment in China soared during the Republic of China. Some 1.5 billion of investment was present in China by the beginning of the 20th century, with Russia, England and Germany being the largest investors. However, with the outbreak of WWI, investment from Germany and

Russia stopped while England and Japan took a leading role. By 1930, foreign investment in China totalled 3.5 billion, with Japan leading (1.4 billion) and England at 1 billion. By 1948, however, the capital stock had halted with investment dropping to only 3 billion, with the US and Britain leading.

### *Currency of the Republic of China*

The currency of China was initially silver-backed, but the nationalist government seized control of private banks in the notorious banking coup of 1935 and replaced the currency with the Fabi, a fiat currency issued by the ROC. Particular effort was made by the ROC government to instill this currency as the monopoly currency of China, stamping out earlier Silver and gold-backed notes that had made up China's currency. Unfortunately, the ROC government used this privilege to issue currency en masse; a total of 1.4 billion Chinese yuan was issued in 1936, but by the end of the second Sino-Japanese war some 1.031 trillion in notes was issued. This trend worsened with the outbreak of the Chinese Civil war in 1946. By 1947, some 33.2 trillion of currency was issued as a result of massive budget deficits resulting from war (taxation revenue was just 0.25 billion, compared with 2500 billion in war expenses). By 1949, the total currency in circulation was 120 billion times more than in 1936..

### *The Chinese War Economy (1937-1945)*

In 1937, Japan invaded China and the resulting warfare literally laid waste to China. Most of the prosperous east China coast was occupied by the Japanese, who carried out various atrocities such as the Rape of Nanjing in 1937 and random massacres of whole villages. In one anti-guerilla sweep in 1942, the Japanese killed up to 200,000 civilians in a month. The war was estimated to have killed between 20 and 25 million Chinese, and destroyed literally all that Chiang had built up in the preceding decade. Development of industries was severely hampered after the war by devastating conflict as well as the inflow of cheap American goods. By 1946, Chinese industries operated at 20% capacity and had 25% of the output of prewar China.

One effect of the war was a massive increase in government control of industries. In 1936, government-owned industries were only 15% of GDP. However, the ROC government took control of many industries in order to fight the war. In 1938, the ROC established a commission for industries and mines to control and supervise firms, as well as instilling price controls. By 1942, 70% of the capital of Chinese industry were owned by the government.

### *Hyperinflation, Civil War and the Relocation of the Republic to Taiwan*

Following the war with Japan, Chiang acquired Taiwan from Japan and renewed his struggle with the communists. However, the corruption of the KMT, as well as hyperinflation as a result of trying to fight the civil war, resulted in mass unrest throughout the Republic and sympathy for the communists. In addition, the communists' promise to redistribute land gained them support among the massive rural population. In 1949, the communists captured Beijing and later Nanjing as well. The People's Republic of China was proclaimed on 1 October 1949. The Republic of China relocated to Taiwan where Japan had laid an educational groundwork. Taiwan continued to prosper under the Republic of China government and came to be known as one of the Four Asian Tigers due to its "economic miracle", and later became one of the largest sources of investment in mainland China after the PRC economy began its rapid growth following Deng's reforms.

### *People's Republic of China (1949 onwards)*

The People's Republic of China is marked by two distinctly different periods: the Mao Era, characterized by a soviet-style planned economy that extinguished the market and eventually became unresponsive, and the post-Mao Era, more properly called the Deng Era (after the reformer who started it) characterized by a transition to a relatively free market economy which was one of the most prosperous periods in China's history. Many Chinese are hopeful that this newfound prosperity will last and China will reclaim her place at the top of nations.

### *The Grand Socialist Experiment: The Mao Era (1949-1976)*

The Mao Era was marked by a soviet-style planned economy which was instituted, despite his promises, in 1949-52. A decade of relatively peaceful development and collectivization followed, but in 1959 Mao launched the disastrous Great Leap forward, an attempt to collectivize all aspects of life (even the peasants' cooking pots), a disaster that was exacerbated by famine. Following this, a reformist faction lead by Deng Xiaoping and Liu Shaoqi forced Mao out of office and experimented with market reforms such as giving peasants private plots. However, these reform efforts were disrupted by Mao's Cultural Revolution, a period of virtually total anarchy and street fighting that resulted in the collapse of the Chinese economy and ended with the execution of the Gang of Four.

### *Initial Policies and Promises of Mao*

Mao initially promised to work with "patriotic capitalists, the petit-bourgeois and other classes" in a period of "New Democracy" in which some capitalism would be allowed. However, in 1952 he broke his promise and accused capitalists of sabotaging China's war effort in Korea. Following this, businesses were collectivized.

### *Collectivization of Industry and Agriculture*

Collectivization of industry took place from 1951 to 1954, by which time industry was entirely in state hands. Despite the communist party's original plans not to finish collectivization of agriculture until 1971, Mao pushed ahead with mutual aid teams by 1953, and People's commune by 1955, forcing the collectivization of agriculture.

### *Economic Developments in the Fifties*

With Soviet aid, Mao set up a basic industrial base that included a small set of industries, mostly related to military matters, that was to become the basis of the later modernization by Deng Xiaoping. New buildings, roads, railroads, and a basic infrastructure was put in place to sustain Chinese development. Illiteracy and a whole host of parasites were eradicated. For these reasons, the 1950s is usually regarded as the high point of the Mao Era.

### *Great Leap Forward*

In 1959, Mao launched a disastrous crash industrialization called the Great Leap forward. This involved making many peasants urban workers, collectivizing virtually everything (including cooking tools) and setting up backyard furnaces to improve steel production. However, the program was a disastrous failure; the new collectivization destroyed the incentive of the peasants, who also overate at people's communes. Most of the new "steel" produced from backyard furnaces was useless. Eventually, famine resulted and an estimated 20 million people died from starvation.

### *Emergence of Reformist Faction*

Reformists, such as Liu Shaoqi and Deng Xiaoping, began to push Mao for reform after the disastrous Great Leap forward. They undid some of Mao's priorities, such as disbanding the People's communes and giving peasants private plots. However, Mao was infuriated and in 1966, he launched a coup against the reformers, calling them Capitalist roader and launching the Cultural Revolution.

### *Cultural Revolution and the Collapse of the Economy*

The Cultural Revolution is usually thought as one of the greatest catastrophes ever to beset China. At Mao's connivance, youth groups, known as red guards, deposed and frequently shot government officials, and then took control of the cities and provinces they ran. This "revolution" began in Shanghai and quickly spread across the country. By 1966, most of the country was in the hands of these "revolutionary committees" who battled each other for power in the cities and provinces they existed.

This disruption, as well as the purge of millions of intellectuals, workers, and officials as "counter-revolutionaries", was disastrous. Economic output fell some 30% over three years, and stagnated for the rest of the period. In addition, an entire generation was deprived of education and hampered China's development for years to come. When referring to this period, Deng Xiaoping said it had created "an entire generation of mental cripples".

### *From Planned Economy to free Market Powerhouse: The Post-Mao Era (1976 Onwards)*

In 1976, at the end of the Cultural Revolution, China's planned economy was in ruins and its people barely surviving.

However, in 1978, China was to witness one of the most rapid periods of change in her 5,000 year history as reformer Deng Xiaoping initiated free market reforms that transformed China's economy. Only 30 years later, China had developed from an economically desolate country ruled by a totalitarian government into an industrial powerhouse, rapidly overtaking developed western nations in recession. The agent of this dramatic change, Deng Xiaoping, is sometimes known as "Deng Gong" or the venerated Deng, for his achievements.

### *Deng Xiaoping's Rise*

An old party stalwart, Deng had been of the reformist faction in the early 1960s and purged in the Cultural Revolution. After Mao's death, however, a coup overthrew the Gang of Four and instilled old party officials such as Deng back in power. However, it was not until 1978 that Hua Guofeng, Mao's appointed successor, resigned and Deng took power, though he was never officially the leader, instead holding the position of chairman of the Central military commission.

### *De-collectivization of Agriculture*

One of Deng's first actions was to break up the People's communes that Mao instated and grant a system of "Bao Chang Dao Hu" (S"u0R7b), in which each plot of land was given to each household to farm. This new, liberalized agriculture was extremely popular, raising grain production tremendously. This system was so successful that in 1983 Deng lifted limits on consumption of many agricultural goods that were instated during the Mao Era due to scarcity. However, these limits were not completely lifted until 1994.

### *Liberalization and Privatization of Business*

Private businesses, which were banned in the Mao Era for being "Capitalist exploiters", were reinstated in the Deng Era. Early on, a citizen could opt to become "Ge Ti Hu" (PSO7b) or self-employed household, and to set up a business instead of taking on state jobs. These "Ge Ti Hu" quickly became extremely wealthy. In the 1990s, many state enterprises were privatized and private individuals were allowed to create companies. In 1990, the Shanghai Stock Exchange was reopened after Mao first closed it 41 years earlier.

Another innovation instated during this period was the Chengbao system or contracting system, in which state assets were given to private operators, who gave the state the money needed for expenses as well as a share of the profits. This system was also rapidly adopted; in the 1980s and 1990s, many schools, hospitals and even bus lines passed from the state to private operators. However, this system was also criticized as many felt that the change in operation for these schools and hospitals, now for-profit, was detrimental to the poor. In addition, some private contractors were accused of gaining their positions solely because of nepotism.

Although privatizations had occurred in the 1980s, it was sped up in the 1990s by Premier Zhu Rongji, who started a policy of privatizing all state enterprises which were losing money. In 1997, the CPC issued a verdict declaring that state-owned companies were now "people-owned companies" who would be subject to mergers and bankruptcy. Thousands of state companies were privatized or partly floated on the stock exchange. In 1978, more than 90% of GDP was produced in state enterprises, which, up to 1992, dominated China's economy. That figure, not accounting for state assets were contracted, had fallen to 30% by 2009.

### *Foreign Investment and Industrialization*

In addition to internal liberalization, Deng also established a series of "special economic zones" in which foreigners could invest in China taking advantage of lower labour costs. This investment helped the Chinese economy boom. In addition, the Chinese government established a series of joint ventures with foreign capital to establish companies in industries hitherto unknown in China. By 2001, China became a member of the World Trade Organization, which has boosted its overall trade in exports/imports—estimated at $851 billion in 2003—by an additional $170 billion a year.

Despite a brief period in 1989 in which foreign capital withdrew from China, China continued to be one of the biggest recipients of foreign investment. In 2006, an estimated $699.5 billion of foreign investment was present in China. A great deal of this investment came from Chinese-speaking regions such as Hong Kong and Taiwan, who was the first to invest in China.

Japanese and Western investment followed. Deng's liberalization of the Chinese economy, along with foreign investment, helped to power China's industrialization. From virtually an industrial backwater in 1978, China is now the world's biggest producer of concrete, steel, ships, textiles as well as the world's biggest auto market. For example, from 2000 to 2006, China's steel production rose from 140 million tons to 416 million tons. From 1975 to 1992, China's auto production rose from 139,800 to 1.1 million automobiles before jumping to 9.35 million in 2008.

### *Developments Post-Deng*

In 1997, Deng Xiaoping died. However, his reformist policies were continued by his successor, Jiang Zemin. The result was a vibrant, growing economy. Under Hu and Wen, who became leaders of China in 2003, the Chinese government continued to give up grounds to private enterprise, yet increased its control in other areas. The new premier, Wen Jiabao, reinstated some Mao-Era social systems, such as social security, as well as sponsoring a new initiative in health care in which the state retook control of hospitals from many contractors who had run them for two decades. However, this was reversed in 2009. Nevertheless, China's economy continued to grow. In 2008, however, it was affected by the global financial meltdown and the growth rate fell to 9.0%. As of 2008, China's GDP (PPP) was between 50 and 60 percent of the GDP (PPP) of the US, while over 10 percent of world GDP (PPP).

To offset the effects of the global economic crisis, the government announced a financial stimulus of around 4 trillion yuan spread over two years. However, new spending by the government was actually only about 1 trillion yuan; the rest was already part of the government's budget.

In mid-2005, China began to experience an enormous property bubble, largely caused by loose monetary policy under premier Wen Jiabao. Property prices tripled from 2005 to 2009, and are continuing to rise.

## Arts, Scholarship and Literature

Chinese characters have had many variants and styles throughout Chinese history. Tens of thousands of ancient written documents are still extant, from oracle bones to Qing edicts. This literary emphasis affected the general perception of cultural refinement in China, e.g. the view that calligraphy was a higher art form than painting or drama. Manuscripts of the Classics and religious texts (mainly Confucian, Taoist, and Buddhist) were handwritten by ink brush.

Calligraphy later became commercialized, and works by famous artists became prized possessions. Chinese literature has a long past; the earliest classic work in Chinese, the *I Ching* or "Book of Changes" dates to around 1000 BC. A flourishing of philosophy during the Warring States Period produced such noteworthy works as Confucius's *Analects* and Laozi's *Tao Te Ching*. (See also: the Chinese classics.) Dynastic histories were often written, beginning with Sima Qian's seminal *Records of the Historian*, which was written from 109 BC to 91 BC.

The Tang Dynasty witnessed a poetic flowering, while the Four Great Classical Novels of Chinese literature were written during the Ming and Qing Dynasties. Printmaking in the form of movable type was developed during the Song Dynasty. Academies of scholars sponsored by the empire were formed to comment on the classics in both printed and handwritten form. Royalty frequently participated in these discussions as well.

The Song Dynasty was also a period of great scientific literature, and saw the creation of works such as Su Song's *Xin Yixiang Fayao* and Shen Kuo's *Dream Pool Essays*. There were also enormous works of historiography and large encyclopedias, such as Sima Guang's *Zizhi Tongjian* of 1084 AD or the *Four Great Books of Song* fully compiled and edited by the 11th century.

For centuries, religious and social advancement in China could be achieved through high performance on the imperial examinations. This led to the creation of a meritocracy, although success was available only to males who could afford test preparation. Imperial examinations required applicants to write essays and demonstrate mastery of the Confucian classics. Those who passed the highest level of the exam became elite scholar-officials known as *jinshi,* a highly esteemed socio-economic position.

Chinese philosophers, writers and poets were highly respected and played key roles in preserving and promoting the culture of the empire. Some classical scholars, however, were noted for their daring depictions of the lives of the common people, often to the displeasure of authorities. The Chinese invented numerous musical instruments, such as the zheng (zither with movable bridges), qin (bridgeless zither), sheng (free reed mouth organ), and xiao (vertical flute) and adopted and developed others such the erhu (alto fiddle or bowed lute) and pipa (pear-shaped plucked lute), many of which later spread throughout East Asia and Southeast Asia, particularly to Japan, Korea, and Vietnam.

### *Demography*

The demographics of China are identified by a large population with a relatively small youth division, which is partially a result of the People's Republic of China's one-child policy. The population policies implemented in China since 1979 have helped to prevent several million births.

Today, China's population is over 1.3 billion, the largest in the world. China plans to conduct its sixth national population census in 2010.

## History

### *Census*

The People's Republic of China conducted censuses in 1953, 1964, and 1982. In 1987 the government announced that the fourth national census would take place in 1990 and that there would be one every ten years thereafter. The 1982 census, which reported a total population of 1,008,180,738, is generally accepted as significantly more reliable, accurate, and thorough than the previous two. Various international organizations eagerly assisted the Chinese in conducting the 1982 census, including the United Nations Fund for Population Activities, which donated US$15.6 million for the preparation and execution of the census.

China has been the world's most populous nation for many centuries. When China took its first post-1949 census in 1953, the population stood at 582 million; by the fifth census in 2000, the population had more than doubled, reaching 1.2 billion.

In the 1920s and 1930s, Chinese interest in social programs through reproductive control, including eugenics, intensified. Beginning in the mid-1950s, the Chinese government introduced, with varying degrees of success, a number of family planning, or population control, campaigns and programs. China's fast-growing population was a major policy matter for its leaders in the mid-twentieth century, so that in the early 1970s, the government implemented the stringent one-child policy (publicly announced in 1979). Under this policy, which had different guidelines for national minorities, married couples were officially permitted only one child. As a result of the policy, China successfully achieved its goal of a more stable and much-reduced fertility rate; in 1971 women had an average of 5.4 children versus an estimated 1.7 children in 2004. Enforcement of the program, however, varied considerably from place to place, depending on the vigilance of local population control workers.

In 1982 China conducted its first population census since 1964. It was by far the most thorough and accurate census taken since 1949 and confirmed that China was a nation of more than 1 billion people, or about one-fifth of the world's population. The census provided demographers with a set of data on China's age-sex structure, fertility and mortality rates, and population density and distribution. Information was also gathered on minority ethnic groups, urban population, and marital status. For the first time since the People's Republic of China was founded, demographers had reliable information on the size and composition of the Chinese work force. The nation began preparing for the 1982 census in late 1976. Chinese census workers were sent to the United States and Japan to study modern census-taking techniques and automation. Computers were installed in every provincial-level unit except Xizang and were connected to a central processing system in the Beijing headquarters of the State Statistical Bureau. Pretests and smallscale trial runs were conducted and checked for accuracy between 1980 and 1981 in twenty-four provincial-level units. Census stations were opened in rural production brigades and urban neighbourhoods. Beginning 1 July 1982, each household sent a representative to a census station to be enumerated. The census required about a month to complete and employed approximately 5 million census takers. The 1982 census collected data in nineteen demographic categories relating to individuals and households. The thirteen areas concerning individuals were name, relationship to head of household, sex, age, nationality, registration status, educational level, profession, occupation, status of nonworking persons, marital status, number of children born and still living, and number of births in 1981. The six items pertaining to households were type (domestic or collective), serial number, number of persons, number of births in 1981, number of deaths in 1981, and number of registered persons absent for more than one year. Information was gathered in a number of important areas for which previous data were either extremely inaccurate or simply nonexistent, including fertility, marital status, urban population, minority ethnic groups, sex composition, age distribution, and employment and unemployment.

A fundamental anomaly in the 1982 statistics was noted by some Western analysts. They pointed out that although the birth and death rates recorded by the census and those recorded through the household registration system were different, the two systems arrived at similar population totals. The discrepancies in the vital rates were the result of the under reporting of both births and deaths to the authorities under the registration system; families would not report some births because of the one-child policy and would not report some deaths so as to hold on to the rations of the deceased.

Nevertheless, the 1982 census was a watershed for both Chinese and world demographics. After an eighteen-year gap, population specialists were given a wealth of reliable, up-to-date figures on which to reconstruct past demographic patterns, measure current population conditions, and predict future population trends. For example, Chinese and foreign demographers used the 1982 census age-sex structure as the base population for forecasting and making assumptions about future fertility trends. The data on age-specific fertility and mortality rates provided the necessary base-line information for making population projections. The census data also were useful for estimating future manpower potential, consumer needs, and utility, energy, and health-service requirements. The sudden abundance of demographic data helped population specialists immeasurably in their efforts to estimate world population. Previously, there had been no accurate information on these 21 percent of the Earth's inhabitants. Demographers who had been conducting research on global population without accurate data on the Chinese fifth of the world's population were particularly thankful for the 1982 breakthrough census.

## *Fertility and Mortality*

In 1949 crude death rates were probably higher than 30 per 1,000, and the average life expectancy was only 32 years. Beginning in the early 1950s, mortality steadily declined; it continued to decline through 1978 and remained relatively constant through 1987. One major fluctuation was reported in a computer reconstruction of China's population trends from 1953 to 1987 produced by the United States Bureau of the Census. The computer model showed that the crude death rate increased dramatically during the famine years associated with the Great Leap Forward (1958–60).

According to Chinese government statistics, the crude birth rate followed five distinct patterns from 1949 to 1982. It remained stable from 1949 to 1954, varied widely from 1955 to 1965, experienced fluctuations between 1966 and 1969, dropped sharply in the late 1970s, and increased from 1980 to 1981. Between 1970 and 1980, the crude birth rate dropped from 36.9 per 1,000 to 17.6 per 1,000. However, elements of socioeconomic change, such as increased employment of women in both urban and rural areas and reduced infant mortality (a greater percentage of surviving children would tend to reduce demand for additional children), may have played some role. The birth rate increased in both 1981 and 1982 to a level of 21 per 1,000, primarily as a result of a marked rise in marriages and first births. The rise was an indication of problems with the one-child policy of 1979. Chinese sources, however, indicated that the birth rate decreased to 17.8 in 1985 and remained relatively constant thereafter.

In urban areas, the housing shortage may have been at least partly responsible for the decreased birth rate. Also, the policy in force during most of the 1960s and the early 1970s of sending large numbers of high school graduates to the countryside deprived cities of a significant proportion of persons of childbearing age and undoubtedly had some effect on birth rates. Primarily for economic reasons, rural birth rates tended to decline less than urban rates. The right to grow and sell agricultural products for personal profit and the lack of an old-age welfare system were incentives for rural people to produce many children, especially sons, for help in the fields and for support in old age. Because of these conditions, it is unclear to what degree education had been able to erode traditional values favouring large families.

Today, the population continues to grow. There is also a serious gender imbalance. Census data obtained in 2000 revealed that 119 boys were born for every 100 girls, and among China's "floating population" the ratio was as high as 128:100. These situations led the government in July 2004 to ban selective abortions of female fetuses. It is estimated that this imbalance will rise until 2025–2030 to reach 20% then slowly decrease.

China now has an increasingly aging population; it is projected that 11.8% of the population in 2020 will be 65 years of age and older. Health care has improved dramatically in China since 1949. Major diseases such as cholera, typhoid, and scarlet fever have been brought under control. Life expectancy has more than doubled, and infant mortality has dropped significantly. On the negative side, the incidence of cancer, cerebrovascular disease, and heart disease has increased to the extent that these have become the leading causes of death. Economic reforms initiated in the late 1970s fundamentally altered methods of providing health care; the collective medical care system has been gradually replaced by a more individual-oriented approach. In Hong Kong, the birth rate of 0.9% is lower than its death rate. Hong Kong's population increases because of immigration from the mainland and a large expatriate population comprising about 4%. Like Hong Kong, Macau also has a low birth rate relying on immigration to maintain its population.

## CIA World Factbook Demographic Statistics

The following demographic statistics are from the CIA World Factbook, unless otherwise indicated. No statistics have been included for areas currently governed by the Republic of China. Unless stated otherwise, statistics refer only to mainland China.

### *Population*

- Mainland only: 1,338,612,968 (2009)
- Hong Kong: 7,055,071 (2009)
- Macau: 559,846 (2009)
- Total: 1,346,227,885 (2009)

### *Historical Population*

- 2100 BC: 14,000,000
- 2 AD: 60,000,000
- 1000: 40,000,000
- 1500: 103,000,000
- 1650: 123,000,000
- 1750: 260,000,000
- 1850: 412,000,000
- 1950: 552,000,000
- 1960: 648,000,000
- 1970: 820,000,000
- 1975: 924,000,000
- 1980: 984,000,000
- 1990: 1,147,000,000
- 2000: 1,264,587,054

### *Population Projection*

- 2000: 1,264,587,054
- 2010: 1,347,000,000
- 2020: 1,430,000,000
- 2030: 1,461,000,000
- 2040: 1,463,144,780
- 2050: 1,465,224,000

### *Population Density*

- National average density: 137.0 people per km$^2$ (2007).

### *Urban-rural Ratio*

- Urban: 42.3% (2007) 562,000,000
- Rural: 57.7% (2007) 767,000,000

### *Age Structure*

- 0–14 years: 19.8% (male 140,877,745/female 124,290,090) (2009)
- 15–64 years: 72.1% (male 495,724,889/female 469,182,087) (2009)
- 65 years and over: 8.1% (male 51,774,115/female 56,764,042) (2009).

### *Further Breakdown of Age Distribution*

- Under 15: 20.3% (2007)
- 15–29: 22.8% (2007)
- 30–44: 26.7% (2007)
- 45–59: 18.2% (2007)
- 60–74: 9.4% (2007)
- 75–84: 2.3% (2007)
- 85 and over: 0.3% (2007).

### *Median Age*

- Total: 34.1 years (2009)
- Male: 33.5 years (2009)
- Female: 34.7 years (2009).

### *Population Growth Rate*

- Population growth rate: 0.606% (2007)
- Natural increase rate: 6.06/1,000 population (2007).

### *Birth Rate*

- Birth rate: 13.45 births/1,000 population (2007).

### *Death Rate*

- Death rate: 7 deaths/1,000 population (2007)

### *Net Migration Rate*

- Net migration rate:-0.39 migrant(s)/1,000 population (2007)

### *Sex Distribution*

- Sex distribution: male 51.53%; female 48.47% (2007)

### *Sex Ratio*

- At birth: 1.11 male(s)/female (2007)

- Under 15 years: 1.134 male(s)/female (2007)
- 15–64 years: 1.057 male(s)/female (2007)
- 65 years and over: 0.914 male(s)/female (2007)
- Total population: 1.06 male(s)/female (2007)

### *Infant Mortality Rate*

- Total: 22.12 deaths/1,000 live births (2007)
- Male: 20.01 deaths/1,000 live births (2007)
- Female: 24.47 deaths/1,000 live births (2007)

### *Child Mortality*

- 415,000 children (under 16) died in China in 2006 (4.3 percent of the world total).

### *Life Expectancy at Birth*

- Total population: 72.88 years (2007)
- Male: 71.13 years (2007)
- Female: 74.82 years (2007)

### *Total Fertility Rate*

- Total fertility rate: 1.75 (avg. births per woman in childbearing years) (2007).

According to the 2000 census, the TFR was 1.22 (0.86 for cities, 1.08 for towns and 1.43 for villages/outposts). Beijing had the lowest TFR at 0.67, while Guizhou had the highest at 2.19. It should be noted that Xiangyang district of Jiamusi city (Heilongjiang) have a TFR of 0.41, which is the lowest TFR recorded anywhere in the world in recorded history. Other extreme low TFR counties are: 0.43 in the Heping district of Tianjin city (Tianjin), and 0.46 in the Mawei district of Fuzhou city (Fujian). At the other end TFR was 3.96 in Geji County (Tibet), 4.07 in Jiali County (Tibet), and 5.47 in Baqing County (Tibet).

### *Marriage and Divorce*

- Marriage rate: 6.3/1,000 population (2006)
- Divorce rate: 1.0/1,000 population (2006).

### *Literacy Rate*

Age 15 and over can read and write:

- Total population: 90.9% (2000 census)

- Male: 95.1% (2000 census)
- Female: 86.5% (2000 census).

***Educational Attainment***

As of 2000, percentage of population age 15 and over having:

- no schooling and incomplete primary: 15.6%
- completed primary: 35.7%
- some secondary: 34.0%
- complete secondary: 11.1%
- some post secondary through advanced degree: 3.6%

***Religious Affiliation***

Only urban population stated (over 1 million people at least), as of 2005:

| | | |
|---|---|---|
| 1. | Shanghai | 10,030,800 |
| 2. | Bèijing | 7,699,300 |
| 3. | Suzhou | 6,521,300 |
| 4. | Tianjin | 4,933,100 |
| 5. | Guangzhou | 4,653,100 |
| 6. | Wuhan | 4,593,400 |
| 7. | Chongqing | 4,239,700 |
| 8. | Shenyang | 3,995,500 |
| 9. | Nanjing | 2,966,000 |
| 10. | Harbin | 2,735,100 |
| 11. | Chengdu | 2,664,000 |
| 12. | Xi'an | 2,657,900 |
| 13. | Jinan | 2,346,000 |
| 14. | Changchun | 2,283,800 |
| 15. | Dalian | 2,181,600 |
| 16. | Hangzhou | 2,059,800 |
| 17. | Shijiazhuang | 1,971,000 |
| 18. | Taiyuan | 1,970,300 |
| 19. | Qingdao | 1,930,200 |
| 20. | Zhengzhou | 1,770,800 |
| 21. | Kunming | 1,597,800 |
| 22. | Lanzhou | 1,576,400 |

23. Changsha 1,562,200
24. Xiamen 1,532,200

### *Households*

- Average household size: 3.1
- Total households: 351,233,698
  - o Of which are family households: 340,491,197 (96.9%)
  - o Of which are collective households: 10,742,501 (3.1%).

### *HIV*

- Adult population (ages 15–49) living with HIV: 0.15% (2008)
- People living with HIV/AIDS: 100,000 (2008)
- HIV/AIDS deaths: 44,000 (2003)

### *Causes of Death*

Major causes of death per 100,000 population, based on 2004 urban population samples:

- malignant neoplasms (cancers): 119.7
- cerebrovascular disease: 88.4
- respiratory diseases: 78.1
- heart diseases: 74.1
- accidents, violence, and poisoning: 43.5

### *Income*

As of 2003, the distribution of urban household income:

- Average per capita disposable income by quintile: Y 9,061 [U.S.$1,095]
  - o first quintile: Y 3,285
  - o second quintile: Y 5,377
  - o third quintile: Y 7,279
  - o fourth quintile: Y 9,763
  - o fifth quintile: Y 17,431

### *Working Life*

Quality of working life:

- Average workweek: 40 hours (1998)

- Annual rate per 100,000 workers for: (1997)
  - injury or accident: 0.7
  - industrial illness: 36
  - death: 1.4
- Death toll from work accidents: 127,000 (2005)
- Funds for pensions and social welfare relief: Y 26,668,000,000 (2001).

### *Access to Services*

- Percentage of population having access to electricity (2000): 98.6%
- Percentage of total population with safe public water supply (2002): 83.6% (urban, rural: 94.0%, 73.0%)
- Sewage system (1999): total (urban, rural)
  - households with flush apparatus 20.7% (50.0%, 4.3%)
  - with pit latrines 69.3% (33.6%, 86.7%)
  - with no latrine 5.3% (7.8%, 4.1%).

### *Social Participation*

- Eligible voters participating in last national election: n/a
- Population participating in voluntary work: n/a
- Trade union membership in total labour force (2005): 18%
- Practicing religious population in total affiliated population: n/a.

### *Social Deviance*

Annual reported arrest rate per 100,000 population (2006) for:

- Property violation: 20.7
- Infringing personal rights: 7.2
- Disruption of social administration: 3.3
- Endangering public security: 1.010.

### *Material Wellbeing*

Urban households possessing (number per household; 2003):

- bicycles: 1.4
- colour televisions: 1.3

- washing machines: 0.9
- refrigerators: 0.9
- cameras: 0.5

Rural families possessing (number per household; 2003):

- bicycles: 1.2
- colour televisions: 0.7
- washing machines: 0.2
- refrigerators: 0.1
- cameras: 0.02

### *Household Income and ExpenditureI*

- Average household size (2005) 3.1; rural households 3.3; urban households 3.0.
- Average annual per capita disposable income of household (2005): rural households Y 3,255 (U.S.$397), urban households Y 10,493 (U.S.$1,281).
- Sources of income (2003): rural households — income from household businesses 75.7%, wages 19.1%, transfers 3.7%, other 1.5%; urban households — wages 70.7%, transfers 23.3%, business income 4.5%, other 1.5%.
- Expenditure: rural (urban) households — food 45.6% (37.1%), housing 15.9% (10.7%), education and recreation 12.1% (14.4%), transportation and communications 8.4% (11.1%), clothing 5.7% (9.8%), medicine and medical service 6.0% (7.1%), household furnishings 4.2% (6.3%).

### *Employment*

- Population economically active (2003): total 760,800,000.
- Activity rate of total population 58.9% (participation rates: over age 15 [2001] 77.7%; female [2001] 37.8%; registered unemployed in urban areas [December 2004] 4.2%).
- Urban employed workforce (2001): 239,400,000; by sector: state enterprises 76,400,000, collectives 28,130,000, self-employment or privately run enterprises 134,870,000.
- Rural employed workforce: 490,850,000.

### *Population Control*

Initially, China's post-1949 leaders were ideologically disposed to view a large population as an asset. But the liabilities of a large, rapidly growing population soon became apparent. For one year, starting in August 1956, vigorous support was given to the Ministry of Public Health's mass birth control efforts. These efforts, however, had little impact on fertility. After the interval of the Great Leap Forward, Chinese leaders again saw rapid population growth as an obstacle to development, and their interest in birth control revived. In the early 1960s, schemes somewhat more muted than during the first campaign, emphasized the virtues of late marriage. Birth control offices were set up in the central government and some provincial-level governments in 1964. The second campaign was particularly successful in the cities, where the birth rate was cut in half during the 1963–66 period. The upheaval of the Cultural Revolution brought the program to a halt, however.

In 1972 and 1973 the party mobilized its resources for a nationwide birth control campaign administered by a group in the State Council. Committees to oversee birth control activities were established at all administrative levels and in various collective enterprises. This extensive and seemingly effective network covered both the rural and the urban population. In urban areas public security headquarters included population control sections. In rural areas the country's "barefoot doctors" distributed information and contraceptives to people's commune members. By 1973 Mao Zedong was personally identified with the family planning movement, signifying a greater leadership commitment to controlled population growth than ever before. Yet until several years after Mao's death in 1976, the leadership was reluctant to put forth directly the rationale that population control was necessary for economic growth and improved living standards.

Population growth targets were set for both administrative units and individual families. In the mid-1970s the maximum recommended family size was two children in cities and three or four in the country. Since 1979 the government has advocated a one-child limit for both rural and urban areas and has generally set a maximum of two children in special circumstances. As of 1986 the policy for minority nationalities was two children per couple, three in special circumstances, and no limit for ethnic groups with very small populations. The overall goal of the one-child policy was to keep the total population within 1.2 billion through the year 2000, on the premise that the Four Modernizations program would be of little value if population growth was not brought under control.

The one-child policy was a highly ambitious population control program. Like previous programs of the 1960s and 1970s, the one-child policy employed a combination of public education, social pressure, and in some cases coercion. The one-child policy was unique, however, in that it linked reproduction with economic cost or benefit.

Under the one-child program, a sophisticated system rewarded those who observed the policy and penalized those who did not. Couples with only one child were given a "one-child certificate" entitling them to such benefits as cash bonuses, longer maternity leave, better child care, and preferential housing assignments. In return, they were required to pledge that they would not have more children. In the countryside, there was great pressure to adhere to the one-child limit. Because the rural population accounted for approximately 60 percent of the total, the effectiveness of the one-child policy in rural areas was considered the key to the success or failure of the program as a whole.

In rural areas the day-to-day work of family planning was done by cadres at the team and brigade levels who were responsible for women's affairs and by health workers. The women's team leader made regular household visits to keep track of the status of each family under her jurisdiction and collected information on which women were using contraceptives, the methods used, and which had become pregnant. She then reported to the brigade women's leader, who documented the information and took it to a monthly meeting of the commune birth-planning committee. According to reports, ceilings or quotas had to be adhered to; to satisfy these cutoffs, unmarried young people were persuaded to postpone marriage, couples without children were advised to "wait their turn," women with unauthorized pregnancies were pressured to have abortions, and those who already had children were urged to use contraception or undergo sterilization. Couples with more than one child were exhorted to be sterilized. The one-child policy enjoyed much greater success in urban than in rural areas. Even without state intervention, there were compelling reasons for urban couples to limit the family to a single child. Raising a child required a significant portion of family income, and in the cities a child did not become an economic asset until he or she entered the work force at age sixteen. Couples with only one child were given preferential treatment in housing allocation. In addition, because city dwellers who were employed in state enterprises received pensions after retirement, the sex of their first child was less important to them than it was to those in rural areas.

Numerous reports surfaced of coercive measures used to achieve the desired results of the one-child policy. The alleged methods ranged from intense psychological pressure to the use of physical force, including some grisly accounts of forced abortions and infanticide. Chinese officials admitted that isolated, uncondoned abuses of the program occurred and that they condemned such acts, but they insisted that the family planning program was administered on a voluntary basis using persuasion and economic measures only. International reaction to the allegations were mixed. The UN Fund for Population Activities and the International Planned Parenthood Association were generally supportive of China's family planning program. The United States Agency for International Development, however, withdrew US$10 million from the Fund in March 1985 based on allegations that coercion had been used.

Observers suggested that an accurate assessment of the one-child program would not be possible until all women who came of childbearing age in the early 1980s passed their fertile years. As of 1987 the one-child program had achieved mixed results. In general, it was very successful in almost all urban areas but less successful in rural areas.

Rapid fertility reduction associated with the one-child policy has potentially negative results. For instance, in the future the elderly might not be able to rely on their children to care for them as they have in the past, leaving the state to assume the expense, which could be considerable. Based on United Nations and Chinese government statistics, it was estimated in 1987 that by the year 2000 the population 60 years and older (the retirement age is 60 in urban areas) would number 127 million, or 10.1 percent of the total population; the projection for 2025 was 234 million elderly, or 16.4 percent. According to projections based on the 1982 census, if the one-child policy were maintained to the year 2000, 25 percent of China's population would be age 65 or older by the year 2040.

## Population Density and Distribution

While China is the most populated country in the world, its national population density (137/km$^2$) is not so high, similar to those of Switzerland and the Czech Republic. The overall population density of PRC conceals major regional variations, the western and northern part have a few million people, while China proper has about 1.2 billion. The vast majority of China's population lives in the fertile plains of the east.

### *Coast and China Proper*

In the 11 provinces, special municipalities, and autonomous regions along the southeast coast, population density was 320.6 people per km$^2$.

Broadly speaking, the population was concentrated in China Proper, east of the mountains and south of the Great Wall. The most densely populated areas included the Yangtze River Valley (of which the delta region was the most populous), Sichuan Basin, North China Plain, Pearl River Delta, and the industrial area around the city of Shenyang in the northeast.

### *Western Areas*

Population is most sparse in the mountainous, desert, and grassland regions of the northwest and southwest. In Inner Mongolia Autonomous Region, portions are completely uninhabited, and only a few sections have populations denser than ten people per km$^2$. The Inner Mongolia, Xinjiang, and Tibet autonomous regions and Qinghai and Gansu provinces comprise 55 percent of the country's land area but in 1985 contained only 5.7 percent of its population.

### *Men/Women Concern*

Future challenges for China will be the gender disparity partially caused by the preference for boys under the 'one-child' system, and the aging of the population, with an increasing problem of young-old disparity. The latter is likely to be tied to the former, as the lack of sufficient female partners for males coming of age is expected to reduce total births.

## Ethnic Groups

The People's Republic of China (PRC) officially recognizes 56 distinct ethnic groups, the largest of which are Han, who constitute about 91.9% of the total population. Ethnic minorities constitute 8.1% or 107.1 million of China's population. Large ethnic minorities include the Zhuang (16 million or 1.30%), Manchu (10 million or 0.86%), Uyghur (9 million or 0.79%), Hui (9 million or 0.79%), Miao (8 million or 0.72%), Yi (7 million or 0.65%), Tujia (5.75 million or 0.62%), Mongols (5 million or 0.47%), Tibetan (5 million or 0.44%), Buyi (3 million or 0.26%), and Korean (2 million or 0.15%).

Ethnic minorities currently experience higher growth rates than the majority Han population, because they are not under the one-child policy. Their proportion of the population in China has grown from 6.1% in 1953, to 8.04% in 1990, 8.41% in 2000 and 9.44% in 2005. Recent surveys indicate that the population growth rate for ethnic minorities is about 7 times greater than that for the Han population. <-citation incorrect

Neither Hong Kong nor Macau recognizes the official ethnic classifications maintained by the central government. In Macau the largest substantial ethnic groups of non-Chinese descent are the Macanese, of mixed Chinese and Portuguese descent (Eurasians), as well as migrants from the Philippines and Thailand. Overseas Filipinas working as domestic workers comprise the largest non-Han Chinese ethnic group in Hong Kong.

### *Languages*

The official spoken standard in the People's Republic of China is Putonghua. Its pronunciation is based on the Beijing dialect of Mandarin

Other languages and dialects include other Mandarin dialects, and Wu (Shanghainese), Yue (Cantonese), Minbei (Fuzhou), Minnan (Hokkien or Taiwanese, Teochiu), Xiang, Gan and Hakka, as well as languages of the minorities. The seven major mutually unintelligible Chinese *dialects*, which are considered by some to be different languages of the Chinese language family, and by some others to be dialects of the Chinese language. Each of these *dialects* has many *sub-dialects*. Over 70% of the Han ethnic group are native speakers of the Mandarin group of dialects spoken in northern and southwestern China. The rest, concentrated in south and southeast China, speak one of the six other major Chinese dialects. In addition to the local dialect, nearly all also speak Standard Chinese or Mandarin (Putonghua), which pronunciation is based on the Beijing dialect, which inself is one of the Mandarin group of dialects, and is the language of instruction in all schools and is used for formal and official purposes.

Non-Chinese languages spoken widely by ethnic minorities include Mongolian, Tibetan, Uyghur and other Turkic languages (in Xinjiang), Korean (in the northeast), and Vietnamese (in the southeast).

In addition to Chinese, in the special administrative regions, English is an official language of Hong Kong and Portuguese is an official language of Macau. *Patua* is a Portuguese creole spoken by a small number of Macanese. English, though not official, is widely used in Macau. In both of the special administrative regions, the dominant spoken form of Chinese is Cantonese.

For written Chinese, the PRC officially uses simplified Chinese characters in mainland China, while traditional Chinese characters are used in Hong Kong and Macau.

The de-facto spoken standard in Hong Kong and Macao is Yue (Cantonese), although officially it is the Chinese language, not specifying, which spoken form is standard. The written official standard in Hong Kong and Macao is in Standard Mandarin in traditional Chinese characters.

On 1 January 1979, the PRC Government officially adopted the hanyu pinyin system for spelling Chinese names and places in mainland China in Roman letters. A system of romanization invented by the Chinese, pinyin has long been widely used in mainland China on street and commercial signs as well as in elementary Chinese textbooks as an aid in learning Chinese characters. Variations of pinyin also are used as the written forms of several minority languages.

Pinyin replaced other conventional spellings in mainland China's English-language publications. The U.S. Government and United Nations also adopted the pinyin system for all names of people and places in mainland China. For example, the capital of the PRC is spelled "Beijing" rather than "Peking".

### *Religions*

The Chinese Communist Government has implemented state atheism since 1949, which makes it difficult to ascertain data on the religious population figures. Thus making the relation between Government and religions was not smooth in the past. But in fact, the people are still holding private worship of traditional religions (Buddhism/Taoism) at home. In recent years, the Chinese government has opened up to religion, especially traditional religions such as Mahayana Buddhism, Taoism and Confucianism because the Government also continued to emphasize the role of religion in building a "Harmonious Society," which was a positive development with regard to the Government's respect for religious freedom.

According to the old Chinese government estimate, there were "over 100 million followers of various faiths" in China. Other estimates put about 100 million or about 8% Chinese who follow Buddhism, with the second largest religion as Taoism (no data), Islam (19 million or 1.5%) and Christianity (14 million or 1%; 4 million Roman Catholics and 10 million Protestants). According to the 1993 edition of The Atlas of Religion, the number of atheists in China is between 10 and 14 percent.

Additionally, the BBC reported in February 2007 that "a poll of 4,500 people by Shanghai university professors found 31.4% of people above the age of 16 considered themselves as religious", a figure that represents 300 million people. Among those surveyed, about 2/3 were "Buddhists, Taoists or worshipers of legendary figures such as the Dragon King and God of Fortune." Other religions represented significantly in that survey were Christianity (40 million) and Islam. China is also known to have small numbers of people who follow Hinduism, Dongbaism, Bon and a number of new religions and sects (particularly Xiantianism and Falun Gong). The official *China Daily* called the Shanghai professors' research "the country's first major survey on religious beliefs". The Chinese government have accepted these new numbers. The wide disparity among these estimates underscores the difficulty of accurately surveying the religious view of a nation of over a billion people and the lack of reliable data.

However, some surveys suggest that the cultural adherents or even outright religious adherents of Buddhism could number as high as 50% to 80% of the population, or about 660 million to over 1 billion. Some estimates for Taoism as high as 400 million or about 30% of the total population, but Adherents.com argues that these are actually numbers for Chinese folk religion or Chinese traditional religion, not Confucianism and Taoism themselves.

The number of adherents to these religions can be overlaid in percentage due to the fact that mostly Chinese consider themselves both Buddhist and Taoist. However, it was difficult to estimate accurately the number of Buddhists because they did not have congregational memberships and often did not participate in public ceremonies.

The minority religions are Christianity (between 40 million, 3%, and 54 million, 4%), Islam (20 million, 1.5%), Hinduism, Dongbaism, Bon and a number of new religions and sects (particularly Xiantianism and Falun Gong).

According to the surveys of Phil Zuckerman on Adherents.com; in 1993, 59% (over 700 million) of the Chinese population was irreligious but in the newest survey (same author) in 2005, it was only 14% (over 180 million).. There are intrinsic logistical difficulties in trying to count the number of religious people anywhere, as well as difficulties peculiar to China. According to Phil Zuckerman, "low response rates," "non-random samples," and "adverse political/cultural climates" are all persistent problems in establishing accurate numbers of religious believers in a given locality. Similar difficulties arise in attempting to subdivide religious people into sects. These issues are especially pertinent in China for two reasons. First, it is a matter of current debate whether several important belief systems in China constitute "religions." As Daniel L. Overmeyer writes, in recent years there has been a "new appreciation...of the religious dimensions of Confucianism, both in its ritual activities and in the inward search for an ultimate source of moral order". Many Chinese belief systems have concepts of a sacred and sometimes spiritual natural world yet do not always invoke a concept of personal god (with the exception of Heaven worship).

The constitution affirms religious toleration subject to several important restrictions. The government places limits on religious practice outside officially recognized organizations. Only two Christian organizations, a Catholic church without ties to the Holy See in Rome and the "Three-Self-Patriotic" Protestant church, are sanctioned by the PRC Government. Unauthorized churches have sprung up in many parts of the country, and unofficial religious practice is flourishing. In some regions authorities have tried to control activities of these unregistered churches. In other regions registered and unregistered groups are treated similarly by authorities, and congregates worship in both types of churches. On 20 July 1999, the Chinese authorities banned and initiated a crackdown on Falun Gong in mainland China.

The Basic Law of Hong Kong protects freedom of religion as a fundamental right. There are a large variety of religious groups in the Hong Kong: Buddhism, Taoism, Confucianism, Christianity including Catholicism, Islam, Hinduism, Sikhism and Judaism all have a considerable number of adherents.

The Macau Basic Law similarly recognizes freedom of religion though the Religious Freedom Ordinance requires registration of religious organizations. The major religions practiced in Macau are Buddhism and traditional beliefs with a smaller minority claiming no religious belief. A small minority of Christians, mostly Catholic, exists.

Hundreds of ethnic groups have existed in China throughout its history. The largest ethnic group in China by far is the Han. This group, however, is internally diverse and can be further divided into smaller ethnic groups that share similar traits.

Over the last three millennia, many previously distinct ethnic groups in China have been Sinicized into a Han identity, which over time dramatically expanded the size of the Han population. However, these assimilations were usually incomplete, and vestiges of indigenous language and culture still often remain in various regions of China. Because of this, many within the Han identity have maintained distinct linguistic and cultural traditions while still identifying as Han. Several ethnicities have also dramatically shaped Han culture, e.g. the Manchurian clothing called the qipao became the new "Chinese" fashion after the 17th century, replacing earlier Han styles of clothing such as the Hanfu. The modern term Chinese nation (*Zhonghua Minzu*) is now used to describe a notion of a Chinese nationality that transcends ethnic divisions.

### *Sports and Recreation*

Many historians believe that association football originated in China, where a form of the sport may have appeared around 1000 AD. Other popular sports include martial arts, table tennis, badminton, and more recently, golf.

Basketball is now popular among young people in urban centers. There are also many traditional sports. Chinese dragon boat racing occurs during the Duan Wu festival. In Inner Mongolia, Mongolian-style wrestling and horse racing are popular. In Tibet, archery and equestrian sports are part of traditional festivals.

Physical fitness is highly regarded. It is common for the elderly to practice Tai Chi Chuan and qigong in parks. Board games such as International Chess, Go (Weiqi), and Xiangqi (Chinese chess) are also common and have organized formal competitions. The capital city of the People's Republic of China, Beijing, hosted the 2008 Olympic Games, a major international sporting event.

### *Science and Technology*

Among the technological accomplishments of ancient China were paper (not papyrus) and papermaking, woodblock printing and movable type printing, the early lodestone and needle compass, gunpowder, toilet paper, early seismological detectors, matches, pound locks, the double-action piston pump, blast furnace and cast iron, the iron plough, the multi-tube seed drill, the suspension bridge, natural gas as fuel, the differential gear for the South Pointing Chariot, the hydraulic-powered armillary sphere, the hydraulic-powered trip hammer, the mechanical chain drive, the mechanical belt drive, the raised-relief map, the propeller, the crossbow, the cannon, the rocket, the multistage rocket, etc.

Chinese astronomers were among the first to record observations of a supernova. The work of the astronomer Shen Kuo (1031–95) alone was most impressive, as he theorized that the sun and moon were spherical, corrected the position of the polestar with his improved sighting tube, discovered the concept of true north, wrote of planetary motions such as retrogradation, and compared the orbital paths of the planets to points on the shape of a rotating willow leaf. With evidence for them, he also postulated geological theories for the processes of land formation in geomorphology and climate change in paleoclimatology.

Other important astronomers included Gan De, Shi Shen, Zhang Heng, Yi Xing, Zhang Sixun, Su Song and Guo Shoujing. Chinese mathematics evolved independently of Greek mathematics and is therefore of great interest in the history of mathematics. The Chinese were also keen on documenting all of their technological achievements, such as in the *Tiangong Kaiwu* encyclopedia written by Song Yingxing (1587–1666).

China's science and technology had fallen behind that of Europe by the 17th century. Political, social and cultural reasons have been given for this, although recent historians focus more on economic causes, such as the high level equilibrium trap. Since the PRC's market reforms, China has become better connected to the global economy and is placing greater emphasis on science and technology.

# 4

# Taiwan

Taiwan, also known as Formosa (from Portuguese: *Ilha Formosa*, "Beautiful Island"), is an island situated in East Asia in the Western Pacific Ocean and located off the southeastern coast of mainland China. Taiwan comprises most of the territory controlled by the Republic of China (ROC) since 1949. "Taiwan" is also the commonly used alternative name both domestically and internationally to refer to the Republic of China.

Separated from the Asian continent by the 120 kilometres (75 mi) wide Taiwan Strait, the main island of the group is 394 kilometres (245 mi) long and 144 kilometres (89 mi) wide. To the northeast are the main islands of Japan and the East China Sea, and the southern end of the Ryukyu Islands of Japan is directly to the east; the Philippines lie to its south across the Bashi Channel. The mountainous island spans the Tropic of Cancer and is covered by tropical and subtropical vegetation. Other minor islands and islets of the group include the Pescadores, Green Island, and Orchid Island as well as the Diaoyutai Islands which have been controlled by Japan since the 1970s and are known as the Senkaku-shoto.

Taiwan has been governed by the Republic of China since 1945 when the ROC acquired Taiwan from Japan as a result of World War II. Four years later the ROC lost the Chinese Civil War to the Communist Party of China and retreated to Taiwan. Taiwan now composes most of ROC's territory and the ROC itself is commonly known as "Taiwan". The political status of Taiwan is disputed because it is claimed by the People's Republic of China (PRC) which was established in 1949 on mainland China and considers itself the successor state to the ROC. Japan had originally acquired Taiwan from the Qing Empire in 1895 per Article 2 of Treaty of Shimonoseki. At the end of World War II, Japan renounced all claims to sovereignty over its former colonial possessions after World War II including Taiwan and Penghu (Pescadores), but did not specify to whom Taiwan and Penghu should be assigned. This fact and subsequent handling of Taiwan's sovereignty by the Allies of World War II led to the complex and unresolved issues of the legal and political status of Taiwan (See below).

Taiwan's rapid economic growth in the decades after World War II has transformed it into an advanced economy as one of the Four Asian Tigers. This economic rise is known as the Taiwan Miracle. It is categorized as an advanced economy by the IMF and high-income economy by the World Bank. Its technology industry plays a key role in the global economy. Taiwanese companies manufacture a large portion of the world's consumer electronics, although most of them are made in their factories in mainland China.

## History

### *Prehistory and Early Settlements*

Evidence of human settlement in Taiwan dates back 30,000 years, although the first inhabitants of Taiwan may have been genetically distinct from any groups currently on the island. About 4,000 years ago, ancestors of current Taiwanese aborigines settled in Taiwan. These aborigines are genetically related to Malay and maternally to Polynesians, and linguists classify their languages as Austronesian. It is thought likely that Polynesian ancestry may be traceable throughout Taiwan.

Records from ancient China indicate that the Han Chinese might have known of the existence of the main island of Taiwan since the Three Kingdoms period (third century, 230 A.D.), having assigned offshore islands in the vicinity names like Greater Liuqiu and Lesser Liuqiu (etymologically, but perhaps not semantically, identical to Ryukyu in Japanese), though none of these names has been definitively matched to the main island of Taiwan. The Ming Dynasty admiral Cheng Ho (Zheng He) visited Taiwan in 1430.

Han Chinese began settling in the Penghu islands in the 1200s, but Taiwan's hostile tribes and its lack of the trade resources valued in that era rendered it unattractive to all but "occasional adventurers or fishermen engaging in barter" until the 16th century.

## European Settlement

In 1544, a Portuguese ship sighted the main island of Taiwan and named it *Ilha Formosa*, which means "Beautiful Island."

In 1624, the Dutch established a commercial base on Taiwan and began to import workers from Fujian and Penghu (Pescadores) as labourers, many of whom settled. The Dutch made Taiwan a colony with its colonial capital at Tayoan City (present day Anping, Tainan). Both *Tayoan* and the island name *Taiwan* derive from a word in Sirayan, one of the Formosan languages.

The Dutch military presence was concentrated at a stronghold called Castle Zeelandia. The Dutch colonists also started to hunt the native Formosan Sika deer (*Cervus nippon taioanus*) that inhabited Taiwan, contributing to the eventual extinction of the subspecies on the island. Furthermore, this contributed to the subsequent identification of native tribes.

In 1626, the Spanish landed on and occupied northern Taiwan (Keelong and Tanshui) as a base to extend its commercial trading. The colonial period lasted 16 years until 1642.

## Koxinga and Qing Rule

Chinese naval and troop forces of Southern Fujian defeated the Dutch in 1662, subsequently expelling the Dutch government and military from the island. They were led by Koxinga. Following the fall of the Ming Dynasty, Koxinga retreated to Taiwan as a self-styled Ming loyalist and established the Kingdom of Tungning (1662–83). Koxinga established his capital at Tainan and he and his heirs, Zheng Jing, who ruled from 1662–82, and Zheng Keshuang, who served less than a year, continued to launch raids on the south-east coast of mainland China well into the Qing Dynasty, attempting to recapture mainland China.

In 1683, following the defeat of Koxinga's grandson by an armada led by Admiral Shi Lang of Southern Fujian, the Qing formally annexed Taiwan, placing it under the jurisdiction of Fujian province. The Qing government tried to reduce piracy and vagrancy in the area, issuing a series of edicts to manage immigration and respect aboriginal land rights. Immigrants mostly from Southern Fujian continued to enter Taiwan. The border between taxpaying lands and "savage" lands shifted eastward, with some aborigines 'Sinicizing' while others retreated into the mountains. During this time, there were a number of conflicts between Chinese from different regions of Southern Fujian, and between Southern Fujian Chinese and aborigines.

Northern Taiwan and the Penghu Islands were the scene of an important subsidiary campaign in the Sino-French War (August 1884 to April 1885). The French occupied Keelung from 1 October 1884 to 22 June 1885 and the Penghu Islands from 31 March to 22 July 1885. A French attempt to capture Tamsui was defeated at the Battle of Tamsui (8 October 1884). Several battles were fought around Keelung between October 1884 and March 1885 between Liu Mingchuan's Army of Northern Taiwan and Colonel Jacques Duchesne's Formosa Expeditionary Corps. The Keelung Campaign, despite some notable French tactical victories, ended in a stalemate. The Pescadores Campaign was a French victory, but had no long-term consequences. The French evacuated both Keelung and the Penghu archipelago at the end of the war.

In 1885, the Qing upgraded Taiwan's status from prefecture of Fujian to full province, the twentieth in the country, with its capital at Taipei. This was accompanied by a modernization drive that included building Taiwan's first railroad and starting a postal service.

### Republic of Formosa & Japanese Rule

Imperial Japan had sought to control Taiwan since 1592, when Toyotomi Hideyoshi began extending Japanese influence overseas. In 1609, the Tokugawa Shogunate sent Arima Harunobu on an exploratory mission. In 1616, Murayama Toan led an unsuccessful invasion of the island. In 1871, an Okinawan vessel shipwrecked on the southern tip of Taiwan and the crew of fifty-four was beheaded by the Paiwan aborigines. The Ryukyu Kingdom kept a tributary relationship with Great Qing at the same time was subordinate to Satsuma Domain of Japan. When Japan sought compensation from Qing China, it was first rejected because Qing considered the incident an internal affair since Taiwan was a prefecture of Fujian Province of Qing and the Ryukyu Kingdom was a tributary of Qing. When Japanese foreign minister Soejima Taneomi asked the compensation again claiming four of the victims were Japanese citizens from Okayama prefecture of Japan, Qing officials rejected the demand on the grounds that the "wild" and "unsubjugated" aboriginals were outside its jurisdiction. Such aboriginals were treated extremely harshly; American consul J.W. Davidson described how the Chinese in Taiwan ate and traded in their aboriginal victims' flesh. The open renunciation of sovereignty led to a Japanese invasion of Taiwan. In 1874, an expeditionary force of three thousand troops was sent to the island. There were about thirty Taiwanese and 543 Japanese casualties (twelve in battle and 531 by endemic diseases for the Japanese side).

Great Qing was defeated in the First Sino-Japanese War (1894–95) and Taiwan and Penghu were ceded in full sovereignty to Japan. Inhabitants wishing to remain Qing subjects were given a two-year grace period to sell their property and move to mainland China. Very few Taiwanese saw this as feasible.

On May 25, 1895, a group of pro-Qing high officials proclaimed the Republic of Formosa to resist impending Japanese rule. Japanese forces entered the capital at Tainan and quelled this resistance on October 21, 1895.

The Japanese were instrumental in the industrialization of the island; they extended the railroads and other transportation networks, built an extensive sanitation system and revised the public school system. During this period, both rice and sugarcane production greatly increased. By 1939, Taiwan was the seventh greatest sugar producer in the world. Still, the Taiwanese and Aborigines were classified as second- and third-class citizens. Large-scale violence continued in the first decade of rule. Japan launched over 160 battles to destroy Taiwan's aboriginal tribes during its 51-year rule of the island...' Around 1935, the Japanese began an island-wide assimilation project to bind the island more firmly to the Japanese Empire and people were taught to see themselves as Japanese. During WWII, tens of thousands of Taiwanese served in the Japanese military. For example, former ROC President Lee Teng-hui's elder brother served in the Japanese navy and died while on duty in the Philippines in February 1945. The Imperial Japanese Navy operated heavily out of Taiwan. The "South Strike Group" was based out of the Taihoku Imperial University in Taiwan. Many of the Japanese forces participating in the Aerial Battle of Taiwan-Okinawa were based in Taiwan. Important Japanese military bases and industrial centers throughout Taiwan, like Kaohsiung, were targets of heavy American bombing.

Japan's rule of Taiwan ended after it lost World War II and signed the Instrument of Surrender of Japan on August 15, 1945. But the Japanese rule had long lasting effects on Taiwan and Taiwanese culture. Japanese pop culture is popular in Taiwan, influenced by the 50-year Japanese rule. Significant parts of Taiwanese infrastructure were started under the Japanese rule. The current Presidential Building was also built during that time. In 1938 there were 309,000 Japanese settlers in Taiwan. After World War II, most of the Japanese were repatriated to Japan.

### *Kuomintang Martial law Period*

The Cairo Conference from November 22 to 26, 1943 in Cairo, Egypt was held to address the Allied position against Japan during WWII, and to make decisions about postwar Asia. One of the three main clauses of the Cairo Declaration was that "all the territories Japan has stolen from China, including Manchuria, Taiwan and the Pescadores, shall be restored to the Republic of China'. However, many challenged that the document was merely a statement of intent for possible reference used for those who would draft the post-war peace treaty and that it was a press release without force of law to transfer sovereignty from Taiwan to the Republic of China. General Order No. 1, issued by Douglas Mac Arthur on September 2, 1945. The general counter-argument for this claim is that while the Cairo Declaration itself was a non-binding declaration, it was given legal effect by the Japanese Instrument of Surrender.

On October 25, 1945, the US Navy ferried ROC troops to Taiwan in order to accept the formal surrender of Japanese military forces in Taipei (then called Taihoku). General Rikichi Ando, governor-general of Taiwan and commander-in-chief of all Japanese forces on the island, signed the instrument of surrender and handed it over to General Chen Yi of the RoC military to complete the official turnover. Chen Yi proclaimed that day to be "Retrocession Day of Taiwan". The ROC administration of Taiwan under Chen Yi was strained by social and political instabilities, which were compounded by economic woes, such as hyperinflation. Furthermore, cultural and linguistic conflicts between Taiwanese and the mainland Chinese quickly led to the loss of popular support for the new government. This culminated in a series of severe clashes between the ROC occupiers and the Taiwanese, in turn leading to the 228 incident and the reign of White Terror.

In 1949, during the Chinese Civil War, the ROC government, led by Chiang Kai-shek, retreated from Nanjing (then romanised as "Nanking") to Taipei, Taiwan's largest city. The ROC continued to claim sovereignty over all "China," which the ROC defines to include mainland China, Taiwan, Outer Mongolia and other areas. The only remaining portions of territory besides Taiwan under ROC control are the Kinmen and Matsu Islands. On the mainland, the victorious Communists established the PRC, claiming to be the sole and only China (which they claimed included Taiwan) and the ROC as a non longer existing entity.

Some 2 million people, consisting mainly of soldiers, Kuomintang party (KMT, Chinese Nationalist Party) members and most importantly the intellectual and business elites were evacuated from mainland China and arrived in Taiwan around that time. In addition, as part of its escape from Communists in mainland China, the ROC government relocated to Taipei with many national treasures including gold reserves and foreign currency reserves. From this period through the 1980s, Taiwan was governed in a state of Martial Law. Little to no distinction was made between the government and the Nationalist party, with public property, government property, and party property being largely interchangeable. Government workers and party members were mostly indistinguishable, with many government workers required to become KMT members, and party workers paid salaries and promised retirement benefits along the lines of government employees. In addition, the creation of other parties were outlawed, and many political opponents were persecuted, incarcerated.

The ROC remained a *de facto* one-party state under martial law under the "Temporary Provisions Effective During the Period of Communist Rebellion", from 1948 to 1987, when the ROC Presidents Chiang Ching-kuo and Lee Teng-hui gradually liberalized and democratized the system. With the advent of democratization, the issue of the political status of Taiwan has resurfaced as a controversial issue (previously, discussion of anything other than unification under the ROC was taboo).

As the Chinese Civil War continued without truce, the ROC built up military fortifications throughout Taiwan. Within this effort, former KMT soldiers built the now famous Central Cross-Island Highway through the Taroko Gorge in the 1950s. The two sides would continue to engage in sporadic military clashes with seldom publicized details well into the 1960s on the nearby islands with an unknown number of night raids. During the Second Taiwan Strait Crisis in September 1958, Taiwan's landscape saw Nike-Hercules missile batteries added, with the formation of the 1st Missile Battalion Chinese Army that would not be deactivated until 1997. Newer generations of missile batteries have since replaced the Nike Hercules systems throughout the island.

During the 1960s and 1970s, Taiwan began to develop into a prosperous, industrialized developed country with a strong and dynamic economy, becoming one of the Four Asian Tigers while maintaining martial law and under the KMT monopoly. Because of the Cold War, most Western nations and the United Nations regarded the ROC as the sole legitimate government of China until the 1970s, when most nations began switching recognition to the PRC.

### *Modern Democratic Era*

Chiang Kai-shek's eventual successor, his son Chiang Ching-kuo, began to liberalize the ROC's political system. In 1984, the younger Chiang selected Lee Teng-hui, an ethnically Taiwanese technocrat, to be his vice president. In 1986, the Democratic Progressive Party (DPP) was formed and inaugurated as the first opposition party in Taiwan to counter the KMT. A year later Chiang Ching-kuo lifted martial law.

After the 1988 death of Chiang Ching-Kuo, President Lee Teng-hui became the first ethnically Taiwanese president of the ROC. Lee continued to democratize the government and decrease the concentration of government authority in the hands of mainland Chinese. Under Lee, Taiwan underwent a process of localization in which Taiwanese culture and history were promoted over a pan-China viewpoint in contrast to earlier KMT policies which had promoted a Chinese identity. Lee's reforms included printing banknotes from the Central Bank rather than the Provincial Bank of Taiwan, and streamlining the Taiwan Provincial Government with most of its functions transferred to the Executive Yuan. Under Lee, the original members of the Legislative Yuan and National Assembly, elected in 1947 to represent mainland Chinese constituencies and having taken the seats without re-election for more than four decades, were forced to resign in 1991. The previously nominal representation in the Legislative Yuan was brought to an end, to reflect the reality that the ROC government had no jurisdiction over mainland China. Restrictions on the use of Taiwanese Hokkien in the broadcast media and in schools were lifted as well. During later years of Lee's administration, he was involved in corruption controversies relating to government release of land and weapons purchase, although no legal proceedings commenced.

In the 1990s, the ROC continued its democratic reforms, as President Lee Teng-hui was elected by the first popular vote held in the ROC during the 1996 Presidential election. In 2000, Chen Shuibian of the DPP, was elected as the first non-KMT President and was re-elected to serve his second and last term since 2004. Polarized politics has emerged in Taiwan with the formation of the Pan-Blue Coalition of parties led by the KMT, favouring eventual Chinese reunification, and the Pan-Green Coalition of parties led by the DPP, favouring an eventual and official declaration of Taiwan independence.

On September 30, 2007, the ruling Democratic Progressive Party approved a resolution asserting separate identity from China and called for the enactment of a new constitution for a "*normal country*". It also called for general use of "*Taiwan*" as the island's name, without abolishing its formal name, the Republic of China. The Chen administration also pushed for referendums on national defence and UN entry in the 2004 and 2008 elections, which failed due to voter turnout below the required legal threshold of 50% of all registered voters. The Chen administration was dogged by public concerns over reduced economic growth, legislative gridlock due to a pan-blue, opposition controlled Legislative Yuan, and corruption involving the First Family as well as government officials.

The KMT increased its majority in the Legislative Yuan in the January 2008 legislative elections, while its nominee Ma Ying-jeou went on to win the presidency in March of the same year, campaigning on a platform of increased economic growth, and better ties with the PRC under a policy of "mutual nondenial". Ma took office on May 20, 2008. Part of the rationale for campaigning for closer economic ties with the PRC stem from the strong economic growth China attained since joining the World Trade Organization. However, some analysts say that despite the election of Ma Ying-jeou, military tensions with the PRC have not been reduced.

## *Geography*

The island of Taiwan lies some 180 kilometers off the southeastern coast of China, across the Taiwan Strait, and has an area of 35,801 km (13,822.8 sq mi). The East China Sea lies to the north, the Philippine Sea to the east, the Luzon Strait directly to the south and the South China Sea to the southwest. The island is characterized by the contrast between the eastern two-thirds, consisting mostly of rugged mountains running in five ranges from the northern to the southern tip of the island, and the flat to gently rolling plains in the west that are also home to most of Taiwan's population. Taiwan's highest point is Yu Shan at 3,952 meters, and there are five other peaks over 3,500 meters. This makes it the world's fourth-highest island. Taroko National Park, located on the mountainous eastern side of the island, has good examples of mountainous terrain, gorges and erosion caused by a swiftly flowing river.

The shape of the main island of Taiwan is similar to a sweet potato seen in a south-to-north direction, and therefore, Taiwanese, especially the Min-nan division, often call themselves "children of the Sweet Potato." There are also other interpretations of the island shape, one of which is a whale in the ocean (the Pacific Ocean) if viewed in a west-to-east direction, which is a common orientation in ancient maps, plotted either by Western explorers or the Great Qing.

## *Geology*

The island of Taiwan lies in a complex tectonic area between the Yangtze Plate to the west and north, the Okinawa Plate on the north-east, and the Philippine Mobile Belt on the east and south. The upper part of the crust on the island is primarily made up of a series of terranes, mostly old island arcs which have been forced together by the collision of the forerunners of the Eurasian Plate and the Philippine Sea Plate. These have been further uplifted as a result of the detachment of a portion of the Eurasian Plate as it was subducted beneath remnants of the Philippine Sea Plate, a process which left the crust under Taiwan more buoyant.

The east and south of Taiwan are a complex system of belts formed by, and part of the zone of, active collision between the North Luzon Trough portion of the Luzon Volcanic Arc and South China, where accreted portions of the Luzon Arc and Luzon forearc form the eastern Coastal Range and parallel inland Longitudinal Valley of Taiwan respectively.

The major seismic faults in Taiwan correspond to the various suture zones between the various terranes. These have produced major quakes throughout the history of the island. On September 21, 1999, a 7.3 quake known as the "921 earthquake" occurred. The seismic hazard map for Taiwan by the USGS shows 9/10 of the island as the highest rating (most hazardous).

On 4 March 2010 at about 01:20 UTC, a magnitude 6.4 earthquake hit southern Taiwan.

## *Climate*

Taiwan's climate is marine tropical. The northern part of the island has a rainy season that lasts from January through late March during the northeast monsoon, and experiences *meiyu* in May. The entire island experiences hot, humid weather from June through September. The middle and southern parts of the island do not have an extended monsoon season during the winter months. Natural hazards such as typhoons and earthquakes are common in the region.

### *Environment and Pollution*

With its high population density and many factories, some areas in Taiwan suffer from heavy pollution. Most notable are the southern suburbs of Taipei and the western stretch from Tainan to Lin Yuan, south of Kaohsiung. In the past, Taipei suffered from extensive vehicle and factory air pollution, but with mandatory use of unleaded gasoline and the establishment of the Environmental Protection Agency, the air quality of Taiwan has improved dramatically. Motor scooters, especially older or cheaper two-stroke versions, which are ubiquitous in Taiwan, also contribute disproportionately to urban air pollution.

### *Natural Resources*

Because of the intensive exploitation throughout Taiwan's pre-modern and modern history, the island's mineral resources (e.g. coal, gold, marble), as well as wild animal reserves (e.g. deer), have been virtually exhausted. Moreover, much of its forestry resources, especially firs were harvested during Japanese rule for the construction of shrines and have only recovered slightly since then. The remaining forests nowadays do not contribute to significant timber production mainly because of concerns about production costs and environmental regulations.

Camphor extraction and sugarcane refining played an important role in Taiwan's exportation from the late nineteenth century through the first half of the twentieth century. The importance of the above industries subsequently declined not because of the exhaustion of related natural resources but mainly of the decline of international market demands.

Nowadays, few natural resources with significant economic value are retained in Taiwan, which are essentially agriculture-associated. Domestic agriculture (rice being the dominant kind of crop) and fisheries retain importance to a certain degree, but they have been greatly challenged by foreign imports since Taiwan's accession to the World Trade Organization in 2001. Consequently, upon the decline of subsistent importance, Taiwan's agriculture now relies heavily on the marketing and exportation of certain kinds of specialty fruits, such as banana, guava, lychee, wax apple, and high-mountain tea.

## *Energy Resources*

Taiwan has significant coal deposits and some insignificant petroleum and natural gas deposits. Electrical power generation is nearly 55% coal-based, 18% nuclear power, 17% natural gas, 5% oil, and 5% from renewable energy sources. Nearly all oil and gas for transportation and power needs must be imported, making Taiwan particularly sensitive to fluctuations in energy prices. Because of this, Taiwan's Executive Yuan is pushing for 10% of energy generation to come from renewable energy by 2010, double from the current figure of approximately 5%. In fact, several wind farms built by American and German companies have come online or will in the near future. Taiwan is rich in wind energy resources, with Wind farms both onshore and offshore, though limited land area favours offshore wind resources. Solar energy is also a potential resource to some extent. By promoting renewable energy, Taiwan's government hopes to also aid the nascent renewable energy manufacturing industry, and develop it into an export market.

## *Ethnic Groups*

Taiwan's population was estimated in 2010 at 23 million, most of whom are on the island of Taiwan. About 98% of the population is of Han Chinese ethnicity. Of these, 86% are descendants of early Han Chinese immigrants known as the "*benshengren*" in Chinese. This group is often referred to "native Taiwanese" in English while the Taiwanese aborigines are also considered as "native Taiwanese" frequently. The *benshengren* group contains two subgroups: the Hoklo people (70% of the total population), whose ancestors migrated from the coastal Southern Fujian (Min-nan) region in the southeast of mainland China starting in the 17th century; and the Hakka (15% of the total population), whose ancestors originally migrated south to Guangdong, its surrounding areas and Taiwan.

12% of population are known as *waishengren* who is composed of people whose ancestors or themselves immigrated from mainland China after the Second World War with the KMT government. The other 2% of Taiwan's population, numbering about 458,000, are listed as the Taiwanese aborigines, divided into 13 major groups: Ami, Atayal, Paiwan, Bunun, Rukai, Puyuma, Tsou, Saisiyat, Tao (Yami), Thao, Kavalan, Truku and Sakizaya.

For sociologists, these ethnic classifications are a social construct, the contestation and compromise between political forces. Sociology scholar Wang Fu-chang writes in his book that Minnanren, Hakka, Waishengren and indigenous peoples are social categories that have developed over the last fifty years.

### Languages

Standard Mandarin is officially recognized by the ROC as the national language and is spoken by the vast majority of residents. About 70% of the people in Taiwan belong to the Hoklo ethnic group and speak both Taiwanese (a variant of Min Nan), as their mother tongue, and Standard Mandarin. Standard Mandarin has been the primary language of instruction in schools since the Japanese were forced out in the 1940s. The Hakka ethnic group, comprising around 15% of the population, use the Hakka language. Taiwan's aboriginal minority groups mostly speak their own native languages, although most also speak Mandarin. The aboriginal languages do not belong to the Chinese or Sino-Tibetan language family, but rather to the Austronesian language family.

Although Mandarin is the language of instruction in schools and dominates television and radio, non-Mandarin languages or dialects have undergone a revival in public life in Taiwan, particularly since the 1990s after restrictions on their use were lifted. A large proportion of the population can speak Taiwanese, and many others have some degree of understanding. People educated during the period of Japanese rule (1895–1945) were taught using Japanese as the medium of instruction. A declining number of persons in the older generations only speak the Japanese they learned in school and the Taiwanese they spoke at home and understand little or no Mandarin.

English is a common second language, with some large private schools providing English instruction. English is compulsory in students' curriculum once they enter elementary school. English as a school subject is also featured on Taiwan's education exams.

## *Religion*

Over 93% of Taiwanese are adherents of a combination of Buddhism, Confucianism, and Taoism; 4.5% are adherents of Christianity, which includes Protestants, Catholics, Latter-day Saints and other, non-denominational, Christian groups;, and 2.5% are adherents of other religions, such as Islam. Taiwanese aborigines comprise a notable subgroup among professing Christians: "...over 64 percent identify as Christian... Church buildings are the most obvious markers of Aboriginal villages, distinguishing them from Taiwanese or Hakka villages."

Confucianism is a philosophy that deals with secular moral ethics, and serves as the foundation of both Chinese and Taiwanese culture. The majority of Taiwanese people usually combine the secular moral teachings of Confucianism with whatever religions they are affiliated with.

One especially important goddess for Taiwanese people is Matsu, who symbolizes the seafaring spirit of Taiwan's ancestors from Fujian and Guangdong.

As of 2009, there are 14,993 temples in Taiwan, approximately one place of worship per 1,500 residents. 9,202 of those temples were dedicated to Taoism. In 2008, Taiwan had 3,262 Churches, an increase of 145.

## *Culture*

The cultures of Taiwan are a hybrid blend of various sources, incorporating elements of traditional Chinese culture, attributable to the historical and ancestry origin of the majority of its current residents, Japanese culture, traditional Confucianist beliefs, and increasingly Western values.

After their move to Taiwan, the Kuomintang imposed an official interpretation of traditional Chinese culture over Taiwanese cultures. The government launched a program promoting Chinese calligraphy, traditional Chinese painting, folk art, and Chinese opera.

Since the Taiwan localization movement of the 1990s, Taiwan's cultural identity has enjoyed greater expression. Identity politics, along with the over one hundred years of political separation from mainland China has led to distinct traditions in many areas, including cuisine and music.

The status of Taiwanese culture is debated. It is disputed whether Taiwanese culture is a regional form of Chinese culture or a distinct culture. Speaking Taiwanese as a symbol of the localization movement has become an emblem of Taiwanese identity.

One of Taiwan's greatest attractions is the National Palace Museum, which houses more than 650,000 pieces of Chinese bronze, jade, calligraphy, painting and porcelain, and is considered one of the greatest collection of Chinese art and objects in the world. The KMT moved this collection from the Forbidden City in Beijing in 1949 when it fled to Taiwan. The collection, estimated to be one-tenth of China's cultural treasures, is so extensive that only 1% is on display at any time. The PRC had said that the collection was stolen and that it legitimately belongs in China, but Taiwan has long defended its collection as a necessary act to protect the pieces from destruction especially during the Cultural Revolution. Relations regarding this treasure have warmed recently as the PRC has agreed to lending relics and that that Beijing Palace Museum Curator Zheng Xinmiao said that artifacts in both Chinese and Taiwanese museums are "China's cultural heritage jointly owned by people across the Taiwan Strait."

Popular sports in Taiwan include basketball and baseball. Taiwan is also the major Asian country for Korfball, where it enjoys mild popularity.

International Community Radio Taipei is the most listened to International Radio Media in Taiwan.Karaoke, drawn from contemporary Japanese culture, is extremely popular in Taiwan, where it is known as KTV. KTV businesses operate in a hotel like style, renting out small rooms and ballrooms varying on the number of guests in a group. Many KTV establishments partner with restaurants and buffets to form all-encompassing elaborate evening affairs for families, friends, or businessmen. Tour buses that travel around Taiwan have several TV's, equipped not for watching movies, but primarily for signing Karaoke. The entertainment counterpart of a KTV is an MTV, being found much less frequently out of the city. There, movies out on DVD can be selected and played in a private theater room. However MTV, more so than KTV, has a growing reputation for being a place that young couples will go to be alone and intimate. Taiwan has a high density of 24-hour convenience stores, which in addition to the usual services, provide services on behalf of financial institutions or government agencies such as collection of parking fees, utility bills, traffic violation fines, and credit card payments. They even provide the service of mailing packages.

Taiwanese culture has also influenced other cultures. Bubble tea and milk tea are available in Australia, Europe and North America. Taiwan television variety shows are very popular in Singapore, Malaysia and other Asian countries. Taiwanese films have won various international awards at film festivals around the world. Ang Lee, a Taiwanese director, has directed critically acclaimed films such as: *Crouching Tiger, Hidden Dragon*; *Eat Drink Man Woman*; *Sense and Sensibility*; *Brokeback Mountain*; and *Lust, Caution*. Other famous Taiwanese directors include Tsai Ming-Liang, Edward Yang and Hou Hsiaohsien.

### *Higher Education*

Taiwan has several major universities, including: National Chengchi University, National Chiao Tung University, National Taiwan University of Science and Technology, National Taipei University, National Taiwan University, National Tsing Hua University, National Cheng Kung University, National Chung Cheng University, National Taiwan Normal University, National Yang Ming University, and National Sun Yat-sen University.

### *Sports*

Baseball is Taiwan's national sport and it is a popular spectator sport. One of the most famous Taiwanese baseball pitchers is Chien-Ming Wang, who is a starting pitcher in Major League Baseball. Other notable players in the league include Chin-hui Tsao who played for the Colorado Rockies (2003–2005) and the Los Angeles Dodgers (2007), Kuo Hong-chih and Hu Chin-lung who are both part of the Los Angeles Dodgers. The Chinese Professional Baseball League in Taiwan was established in 1989, and eventually absorbed the competing Taiwan Major League in 2003. As of 2008, the CPBL has four teams with average attendance of approximately 3,000 per game.

Besides baseball, martial arts such as taekwondo, karate, and kung fu are also widely practiced and competed.

In 2009, Taiwan hosted two international sporting events on the island. The World Games 2009 were held in Kaohsiung City between July 16, 2009 and July 26, 2009. Taipei City hosted the 21st Summer Deaflympics in September of the same year.

## *Political Status*

The controversy regarding the political status of Taiwan hinges on whether Taiwan, Penghu, Kinmen, and Matsu should remain effectively independent as territory of the Republic of China (ROC, a distinct yet related political entity to the People's Republic of China), become unified with the territories now governed by the People's Republic of China (PRC), or formally declare independence and become the Republic of Taiwan. The controversy over the political status of the Republic of China hinges on whether its existence as a state is legitimate and recognized by the international community.

Currently, Taiwan, Penghu, Kinmen, Matsu and some other minor islands effectively make up the jurisdiction of the state with the official name of Republic of China, known in the West as Taiwan. The ROC, which took control of Taiwan (including Penghu and other nearby islands) in 1945, ruled mainland China and claimed sovereignty over Outer Mongolia and Tannu Uriankhai (part of which is present day Tuva) before losing the Chinese Civil War and relocating its government to Taipei in December 1949.

Since the ROC lost its United Nations seat as "China"in 1971 (replaced by the PRC), most sovereign states have switched their diplomatic recognition to the PRC, recognizing or acknowledging the PRC to be the sole legitimate representative of all China, though many deliberately avoid stating clearly what territories they believe China includes. As of January 2008, the ROC maintains official diplomatic relations with 23 sovereign states, although *de facto* relations are maintained with nearly all others. Agencies such as the Taipei Economic and Cultural Representative Office and American Institute in Taiwan operate as *de facto* embassies without official diplomatic status.

The ROC government has in the past actively pursued the claim as the sole legitimate government over mainland China and Taiwan. This position started to be largely adjusted in the early 1990s as democracy was introduced and new Taiwanese leaders were elected, changing to one that does not actively challenge the legitimacy of PRC rule over mainland China. However, with the election of the Kuomintang (KMT, "Chinese Nationalist Party") back into executive power in 2008, the ROC government has changed its position back to that "mainland China is also part of the territory of the ROC." Both the PRC and the ROC carry out cross-strait relations through specialized agencies (such as the Mainland Affairs Council of the ROC), rather than through foreign ministries. Different groups have different concepts of what the current formal political situation of Taiwan is. In addition, the situation can be confusing because of the different parties and the effort by many groups to deal with the controversy through a policy of deliberate ambiguity. The political solution that is accepted by many of the current groups is the following perspective of the status quo: that is, to unofficially treat Taiwan as a state and at a minimum, to officially declare no support for the government of this state making a formal declaration of independence. What a formal declaration of independence would consist of is not clear and can be confusing given the fact that the People's Republic of China has never controlled Taiwan since its founding and the fact that the Republic of China, whose government controls Taiwan, is still a sovereign state as established at Nanjing in 1911, but its territory limited to the province of Taiwan, Penghu, a part of the Nansha islands and Diaoyu island. The status quo is accepted in large part because it does not define the legal status or future status of Taiwan, leaving each group to interpret the situation in a way that is politically acceptable to its members. At the same time, a policy of status quo has been criticized as being dangerous precisely because different sides have different interpretations of what the status quo is, leading to the possibility of war through brinkmanship or miscalculation.

### *Background*

Taiwan (excluding Penghu) was first populated by Austronesian people and was colonized by the Dutch, who had arrived in 1623. The Kingdom of Tungning, lasting from 1661 to 1683, was the first Han Chinese government to rule Taiwan. From 1683, the Qing Dynasty ruled Taiwan as a prefecture and in 1875 divided the island into two prefectures. In 1885 the island was made into a separate Chinese province to speed up development in this region. In the aftermath of the First Sino-Japanese War, Taiwan and Penghu were ceded by the Qing Dynasty to Japan in 1895. Japanese troops in Taiwan surrendered to the Republic of China at end of World War II, putting Taiwan under a Chinese government again after 50 years of Japanese rule. The ROC would then claim sovereignty on the basis of the Qing dynasty's administration, Cairo Declaration, Potsdam Declaration, and Japanese Instrument of Surrender, but this became contested by pro-independence groups in subsequent years due to different perceptions of the said documents' legality. Upon losing the Chinese civil war in 1949, the ROC government retreated to Taipei, and kept control over a few islands along the coast of mainland China and in the South China Sea. The People's Republic of China (PRC) was established in mainland China on October 1, 1949, claiming to be the successor to the ROC.

Quemoy, Matsu and Wuchiu on the coast of Fukien, and Taiping and Pratas in the South China Sea, are part of the ROC's present territory, but were not ceded to Japan. Some arguments supporting the independence of Taiwan do not apply to these islands.

## Question of Sovereignty over Taiwan

### *Cession, Retrocession and Self-determination of Taiwan*

China, during the Qing Dynasty, ceded the island of Taiwan, including Penghu, to Japan "in perpetuity" at the end of the First Sino-Japanese War by signing the Treaty of Shimonoseki. In the Cairo Conference of 1943, the allied powers agreed to have Japan restore "all the territories Japan has stolen from the Chinese", specifically listing "Formosa" and Penghu, to the Republic of China after the defeat of Japan. According to both the People's Republic of China and the Republic of China, this agreement was given legal force by the Instrument of Surrender of Japan in 1945. The PRC's UN Ambassador, Wang Yingfan, has stated multiple times in the UN general committee: "Taiwan is an inseparable part of China's territory since antiquity" and "both the 1943 Cairo Declaration and the 1945 Potsdam Declaration have reaffirmed in unequivocal terms China's sovereignty over Taiwan as a matter of international law." The PRC rejects arguments involving the lack of a specific treaty (San Francisco Peace Treaty) transferring Taiwan's sovereignty to China by noting that it was not a signatory to any such treaty, making the treaties irrelevant with regard to Chinese claims. The ROC argues that the Treaty of Taipei implicitly transferred sovereignty of Taiwan to it, however the US State Dept. denied such an interpretation in its 1971 Starr Memorandum

On the other hand, a number of supporters of Taiwan independence argue that Taiwan was only formally incorporated as a Chinese territory under the Qing Dynasty in 1683, and as a province in 1885. Subsequently, because of the Shimonoseki Treaty of 1895, Taiwan had been *de jure* part of Japan when the ROC was established in 1912 and thus was not part of the Chinese republic. Also, because the Cairo Declaration was an unsigned press communique, the independence advocates argue that the legal effectiveness of the Declaration is highly questionable. Furthermore, they point out that the Instrument of Surrender of Japan was no more than an armistice, a "modus vivendi" in nature, which served as a temporary or provisional agreement that would be replaced with a peace treaty. Therefore, only a military occupation of Taiwan began on Oct. 25, 1945, and both the Treaty of San Francisco and Treaty of Taipei hold legal supremacy over the surrender instrument. These treaties did not transfer the title of Taiwan from Japan to China. According to this argument, the sovereignty of Taiwan was returned to the people of Taiwan when Japan renounced sovereignty of Taiwan in the Treaty of San Francisco (also known as San Francisco Peace Treaty, SFPT) in 1951, based on the policy of self-determination which has been applied to "territories which detached from enemy states as a result of the Second World War" as defined by article 76b and 77b of the United Nations Charter and also by the protocol of the Yalta Conference. The United Nations General Assembly has not been particularly receptive to this argument, and the ROC's applications for admission to the United Nations have been rejected over 16 times since the early 1990s. Although the interpretation of the peace treaties was used to challenge the legitimacy of the ROC on Taiwan before the 1990s, the introduction of popular elections in Taiwan has compromised this position. Except for the most extreme Taiwan independence supporters, most Taiwanese support the popular sovereignty theory and no longer see much conflict between this theory of sovereignty and the ROC position. In this sense, the ROC government currently administering Taiwan is not the same ROC which accepted Japanese surrender because the ruling authorities were given popular mandate by different pools of constituencies: one is the mainland Chinese electorate, the other is the Taiwanese constituencies. In fact, former president Chen Shuibian has been frequently emphasizing the popular sovereignty theory in his speeches.

However, as of 2007, the conflict between these two theories still plays a role in internal Taiwanese politics. The popular sovereignty theory, which the pan-green coalition emphasizes, suggests that Taiwan could make fundamental constitutional changes by means of a popular referendum. The ROC legal theory, which is supported by the pan-blue coalition, suggests that any fundamental constitutional changes would require that the amendment procedure of the ROC constitution be followed.

### *Position of the People's Republic of China (PRC)*

The position of the PRC is that the ROC ceased to be a legitimate government upon the founding of the former on October 1, 1949 and that the PRC is the successor of the ROC as the sole legitimate government of China, with the right to rule Taiwan under the succession of states theory.

The ROC argues that it maintains all the characteristics of a state and that it was not "replaced" or "succeeded" by the PRC because it has continued to exist long after the PRC's founding. According to the Montevideo Convention of 1933, the most cited source for the definition of statehood, a state must possess a permanent population, a defined territory, a government, and the capacity to enter into relations with other states. The ROC claims to meet all these criteria as it possesses a government exercising effective jurisdiction over well-defined territories with over 23 million permanent residents and a full fledged foreign ministry.

The PRC argues that the ROC and PRC are two different factions in the Chinese Civil War, which never legally ended. Therefore both factions belong to the same sovereign country—China. Since Taiwan's sovereignty belongs to China, the secession of Taiwan should be agreed upon by 1.3 billion Chinese citizens instead of the 23 million ROC citizens who currently live in Taiwan. Furthermore, they interpret UN General Assembly Resolution 2758, which states "Recognizing that the representatives of the Government of the People's Republic of China are the only lawful representatives of China to the United Nations", to mean that the PRC is recognized as having the sovereignty of all of China, including Taiwan. The actual Resolution, however, does not specifically mention Taiwan and to whom its sovereignty belongs. Therefore, the PRC believes that it is within their legal rights to extend its jurisdiction to Taiwan, by military means if necessary.

In addition, the PRC argues that the ROC does not meet the fourth criterion of the Montevideo Convention, as it is recognized by only 23 states and has been denied access to international organizations such as the UN. The ROC counters that the pressure the PRC exerts prevents the ROC from being widely recognized. This was accomplished because the PRC took many coercive steps to isolate the ROC diplomatically. Furthermore, the ROC points out that Article 3 of the same Montevideo Convention specifically says, "The political existence of the state is independent of recognition by other states." Nevertheless, the PRC points out the fact that the Montevideo Convention was only signed by 19 states at the Seventh International Conference of American States. Thus the authority of the United Nations as well as UN Resolutions should supersede the Montevideo Convention.

The People's Republic of China until the 1990s had made it clear that "there is only one China in the world", "Taiwan is an inalienable part of China" and "the Government of the People's Republic of China is the sole legal government of China". This view was rejected by the Republic of China government which held the view that both PRC and ROC were two separate and sovereign Chinese governments that split during the Chinese civil war. The current position of the People's Republic of China is more ambiguous on its terms of reunification. It is clear that the PRC still maintains that "there is only one China in the world" and "Taiwan is an inalienable part of China", however instead of "the Government of the People's Republic of China is the sole legal government of China", the PRC now emphasizes that "both Taiwan and the mainland belong to one and the same China". Although the current position allows for flexibility in terms of defining that "one China", any departure from the One-China policy is deemed unacceptable by the PRC government. The PRC government is unwilling to negotiate with the Republic of China government under any formulation other than under a One-China policy, although a more flexible definition of "one China" such as found in the 1992 consensus is possible under PRC policy. The PRC government considers the 1992 consensus a temporary measure to set aside sovereignty disputes and to enable talks.

The Republic of China government under the DPP administration (2000-May 2008) was unwilling to accept the 1992 consensus as it or any formulation under a "one-China policy" contradicts with the DPP position of "one country on each side"-Taiwan and China are two different and unrelated countries on each side of the Taiwan strait. Since the Presidential election victory of KMT, the Republic of China government has accepted the 1992 consensus, which results in semiofficial talks between the two sides. The ROC recently interprets the cross-strait relations as "special", "but not that between two nations". However, it remains to be seen how the PRC evolves from its current position of a more ambiguous definition on "one China" when it comes to ultimately resolving the sovereignty disputes between the two sides: whether "China" will mean the PRC or a new entity that includes the current PRC and ROC.

The PRC government considers perceived violations of its One-China policy, or inconsistencies such as supplying the ROC with arms, a violation of its rights to territorial integrity. International news organizations often report that "China considers Taiwan a renegade province that must be united with the mainland by force if necessary", even though the PRC does not explicitly say that Taiwan is a renegade province.

However, official PRC media outlets and officials often refer to Taiwan as "China's Taiwan Province". (Note: the PRC also claims Quemoy, Wuchiu and Matsu as part of its Fujian Province, and the South China Sea Islands part of its Guangdong and Hainan provinces.)

### *Position of the Republic of China (ROC)*

Both the original 1912 constitution and the 1923 'Cao' version failed to list Taiwan as a part of the ROC since the framers at the time considered Taiwan to be Japanese territory. It was only in the mid 1930's when both the CCP and KMT realised the future strategic importance of Taiwan that they altered their party positions to make a claim on Taiwan as a part of China. After losing the Civil War against the Communist Party in 1949, Chiang Kai Shek (Jiang JieShi) and the Nationalist Party fled to Taiwan, establishing a new government there, but always maintained that their government represented all of China, i.e. both Taiwan and the mainland.

The position of most supporters of Taiwan independence is that the PRC is the government of China, that Taiwan is not part of China, and the 'Republic of China (Taiwan)' is an independent, sovereign state. The Democratic Progressive Party states that Taiwan has never been under the jurisdiction of the PRC, and that the PRC does not exercise any hold over the 23 million Taiwanese on the island. On the other hand, the position of most Chinese reunification supporters believe the Chinese Civil War is still not concluded as no peace agreement was even signed. Therefore, the current political separation across the Taiwan strait is only temporary and a reunified China including both mainland China and Taiwan will be the result.

The position of the Republic of China had been that it was a *de jure* sovereign state. "Republic of China," according to the ROC government's definition, extended to both mainland China and the island of Taiwan.

In 1991, President Lee Teng-hui unofficially claimed that the government would no longer challenge the rule of the Communists in mainland China, the ROC government under Kuomintang (KMT) rule actively maintained that it was the sole legitimate government of China. The Courts in Taiwan have never accepted President Lee's statement, primarily due to the reason that the (now defunct) National Assembly never officially changed the national borders. Notably, the People's Republic of China claims that changing the national borders would be "a precursor to Taiwan independence". The task of changing the national borders now requires a constitutional amendment passed by the Legislative Yuan and ratified by a majority of all eligible ROC voters, which the PRC has implied would constitute grounds for military attack.

On the other hand, though the constitution of the Republic of China promulgated in 1946 does not state exactly what territory it includes, the draft of the constitution of 1925 did individually list the provinces of the Republic of China and Taiwan was not among them, since Taiwan was arguably *de jure* part of Japan as the result of the Treaty of Shimonoseki of 1895.

The constitution also stipulated in Article I.4, that "the territory of the ROC is the original territory governed by it; unless authorized by the National Assembly, it cannot be altered." However, in 1946, Sun Fo, son of Sun Yat-Sen and the minister of the Executive Yuan of the ROC, reported to the National Assembly that "there are two types of territory changes: Renouncing territory and annexing new territory. The first example would be the independence of Mongolia, and the second example would be the reclamation of Taiwan. Both would be examples of territory changes." Japan renounced all rights to Taiwan in the Treaty of San Francisco in 1951 and the Treaty of Taipei of 1952 without an explicit recipient.

While the ROC continuously ruled Taiwan after the government was directed to Taiwan by the General Order No. 1 (1945) to receive Japanese surrender, there has never been a meeting of the ROC National Assembly in making a territory change according to the ROC constitution. The explanatory memorandum to the constitution explained the omission of individually listing the provinces as opposed to the earlier drafts was an act of deliberate ambiguity: as the ROC government does not recognize the validity of the Treaty of Shimonoseki, based on Chiang Kai-shek's Denunciation of the treaty in the late 1930s, hence (according to this argument) the sovereignty of Taiwan was never disposed by China. A ratification by the ROC National Assembly is therefore unnecessary. Significantly, the United States denied such an interpretation with its ratification of the Washington Naval Treaty of 1922, which specifically recognized Formosa and the Pescadores as an insular area of Japan.

The Additional Articles of the Constitution of the Republic of China have mentioned "Taiwan Province," and the now defunct National Assembly passed constitutional amendments that give the people of the "Free Area of the Republic of China", comprising the territories under its current jurisdiction, the sole right, until reunification, to exercise the sovereignty of the Republic through elections of the President and the entire Legislature as well as through elections to ratify amendments to the ROC constitution. Also, Chapter I, Article 2 of the ROC constitution states that "The sovereignty of the Republic of China shall reside in the whole body of citizens." This suggests that the constitution implicitly admits that the sovereignty of the ROC is limited to the areas that it controls even if there is no constitutional amendment that explicitly spells out the ROC's borders.

In 1999, ROC President Lee Teng-hui proposed a two-state theory in which both the Republic of China and the People's Republic of China would acknowledge that they are two separate countries with a special diplomatic, cultural and historic relationship. This however drew an angry reaction from the PRC who believed that Lee was covertly supporting Taiwan independence.

President Chen Shuibian (2000-May 2008) fully supported the idea that the "Republic of China is an independent, sovereign country" but held the view that the Republic of China is Taiwan and Taiwan does not belong to the People's Republic of China. This is suggested in his Four-stage Theory of the Republic of China. Due to the necessity of avoiding war with the PRC however, President Chen had refrained from formally declaring Taiwan's independence. Government publications have implied that Taiwan refers to the ROC, and "China" refers to the PRC. After becoming chairman of the Democratic Progressive Party in July 2002, Chen appeared to move further than Lee's special two-state theory and in early August 2002, by putting forward the "one country on each side" concept, he stated that Taiwan may "go on its own Taiwanese road" and that "it is clear that the two sides of the straits are separate countries." These statements essentially eliminate any "special" factors in the relations and were strongly criticized by opposition parties in Taiwan. President Chen has repeatedly refused to endorse the One China Principle or the more "flexible" 1992 Consensus the PRC demands as a precursor to negotiations with the PRC. During Chen's presidency, there had not been any successful attempts to restart negotiations on a semiofficial level.

In the 2008 ROC elections, the people delivered KMT's Ma Ying-jeou with an election win as well as a sizable majority in the legislature. President Ma, throughout his election campaign, maintained that he would accept the 1992 consensus and promote better relations with the PRC. In respect of Taiwan political status, his policy was he would not negotiate with the PRC on the subject of reunification during his term; he would never declare Taiwan independence; and he would not provoke the PRC into attacking Taiwan. He officially accepted the 1992 Consensus in his inauguration speech which resulted in direct semiofficial talks with the PRC, and this later led to the commencement of weekend direct charter flights between mainland China and Taiwan. President Ma also interprets the cross-strait relations as "special", "but not that between two nations". He later stated that mainland China is part of the territory of the Republic of China, and laws relating to international relations are not applicable to the relations between mainland China and Taiwan, as they are parts of a state.

## *Position of other Countries and International Organizations*

Because of anti-communist sentiment at the start of the Cold War, the Republic of China was initially recognized as the sole legitimate government of China by the United Nations and most Western nations. On January 9, 1950, the Israeli government extended recognition to the People's Republic of China. United Nations General Assembly Resolution 505, passed on February 1, 1952 considered the Chinese communists to be rebels against the Republic of China. However, the 1970s saw a switch in diplomatic recognitions from the ROC to the PRC. On 25 October 1971, Resolution 2758 was passed by the UN

General Assembly, which "decides to restore all its rights to the People's Republic of China and to recognize the representatives of its Government as the only legitimate representatives of China to the United Nations, and to expel forthwith the representatives of Chiang Kai-shek from the place which they unlawfully occupy at the United Nations and in all the organizations related to it." Multiple attempts by the Republic of China to rejoin the UN, no longer to represent all of China but just the people of the territories it governs, have not made it past committee, largely due to diplomatic maneuvering by the PRC, which claims Resolution 2758 has settled the matter.

The PRC refuses to maintain diplomatic relations with any nation that recognizes the ROC, but does not object to nations conducting economic, cultural, and other such exchanges with Taiwan that do not imply diplomatic relation. Therefore, many nations that have diplomatic relations with Beijing maintain quasi-diplomatic offices in Taipei. For example, the United States maintains the American Institute in Taiwan. Similarly, the government in Taiwan maintains quasi-diplomatic offices in most nations under various names, most commonly as the Taipei Economic and Cultural Office.

The United States, Canada, the United Kingdom, Republic of India, Pakistan and Japan have formally adopted the One China policy, under which the People's Republic of China is theoretically the sole legitimate government of China. However, the United States and Japan *acknowledge* rather than *recognize* the PRC position that Taiwan is part of China. In the case of Canada and the UK, bilateral written agreements state that the two respective parties *take note* of Beijing's position but do not use the word *support*. The UK government position that "the future of Taiwan be decided peacefully by the peoples of both sides of the Strait" has been stated several times. Despite the PRC claim that the United States opposes Taiwanese independence, the United States takes advantage of the subtle difference between "oppose" and "does not support". In fact, a substantial majority of the statements Washington has made says that it "does not support Taiwan independence" instead of saying that it "opposes" independence. Thus, the US currently does not take a position on the political outcome, except for one explicit condition that there be a peaceful resolution to the differences between the two sides of the Taiwan Strait. All of this ambiguity has resulted in the United States constantly walking on a diplomatic tightrope with regard to cross strait relations.

The ROC maintains formal diplomatic relations with 23 countries, mostly in Central America and Africa. Notably the Holy See also recognizes the ROC, a largely non-Christian/ Catholic state, mainly to protest what it sees as the PRC's suppression of the Catholic faith in mainland China. However, Vatican diplomats were engaged in talks with PRC politicians at the time of Pope John Paul II's death, with a view towards improving relations between the two countries. When asked, one Vatican diplomat suggested that relations with Taiwan might prove "expendable" should PRC be willing to engage in positive diplomatic relations with the Holy See. Under Pope Benedict XVI the Vatican and PRC have shown greater interest in establishing ties, including the appointment of pro-Vatican bishops and the Pope canceling a planned visit from the Dalai Lama.

During the 1990s, there was a diplomatic tug of war in which the PRC and ROC attempted to outbid each other to obtain the diplomatic support of small nations. This struggle seems to have slowed as a result of the PRC's growing economic power and doubts in Taiwan as to whether this aid was actually in the Republic of China's interest. In March 2004, Dominica switched recognition to the PRC in exchange for a large aid package. However, in late 2004, Vanuatu briefly switched recognition from Beijing to Taipei, leading to the ousting of its Prime Minister and a return to its recognition of Beijing. On January 20, 2005, Grenada switched its recognition from Taipei to Beijing, in return for millions in aid (US$1,500 for every Grenadian). However, on May 14, 2005, Nauru announced the restoration of formal diplomatic relations with Taipei after a three-year hiatus, during which it briefly recognized the People's Republic of China.

On October 26, 2005, Senegal broke off relations with the Republic of China and established diplomatic contacts with Beijing. The following year, on August 5, 2006, Taipei ended relations with Chad when Chad established relations with

Beijing. On April 26, 2007, however, Saint Lucia, which had previously severed ties with the Republic of China following a change of government in December 1996, announced the restoration of formal diplomatic relations with Taipei. On June 7, 2007, Costa Rica broke off diplomatic ties with the Republic of China in favour of the People's Republic of China. In January 2008 Malawi's foreign minister reported Malawi decided to cut diplomatic recognition of the Republic of China and recognize the People's Republic of China.

Under continuing pressure from the PRC to bar any representation of the ROC that may imply statehood, international organizations have adopted different policies toward the issue of ROC's participation. In cases where almost all UN members or sovereign states participate, such as the World Health Organization, the ROC has been completely shut out, while in others, such as the World Trade Organization (WTO) and International Olympic Committee (IOC) the ROC participates under unusual names: "Chinese Taipei" in the case of APEC and the IOC, and the "Separate Customs Territory of Taiwan, Penghu, Kimmen and Matsu" (often shortened as "Chinese Taipei") in the case of the WTO. The issue of ROC's name came under scrutiny during the 2006 World Baseball Classic. The organizers of the 16-team tournament intended to call Taiwan as such, but reverted to "Chinese Taipei" under pressure from PRC. The ROC protested the decision, claiming that the WBC is not an IOC event, but did not prevail. The ISO 3166 directory of names of countries and territories registers Taiwan (TW) separately from and in addition to the People's Republic of China (CN), but lists Taiwan as "Taiwan, Province of China" based on the name used by the UN under PRC pressure. In ISO 3166-2:CN, Taiwan is also coded CN-71 under China, thus making Taiwan part of China in ISO 3166-1 and ISO 3166-2 categories. Naming issues surrounding Taiwan/ROC continue to be a contentious issue in non-governmental organizations such as the Lions Club, which faced considerable controversy naming its Taiwanese branch.

### *Controversies*

Many political leaders who have maintained some form of One-China Policy have committed slips of the tongue in referring to Taiwan as a country or as the Republic of China. United States presidents Ronald Reagan and George W. Bush have been known to have referred to Taiwan as a country during their terms of office. Although near the end of his term as U.S. Secretary of State, Colin Powell said that Taiwan is not a state, he referred to Taiwan as the Republic of China twice during a testimony to the U.S. Senate Foreign Relations Committee on March 9, 2001. In the People's Republic of China Premier Zhu Rongji's farewell speech to the National People's Congress, Zhu accidentally referred to Mainland China and Taiwan as two countries. There are also those from the PRC who informally refer to Taiwan as a country. South Africa delegates once referred to Taiwan as the "Republic of Taiwan" during Lee Teng-hui's term as President of the ROC. In 2002, Michael Bloomberg, the mayor of New York City, referred to Taiwan as a country. Most recently, former US Secretary of Defence Donald Rumsfeld stated in a local Chinese newspaper in California in July 2005 that Taiwan is "a sovereign nation". The People's Republic of China discovered the statement about three months after it was made.

In a controversial speech on February 4, 2006, Japanese Foreign Minister Taro Aso called Taiwan a country with very high education levels because of previous Japanese colonial rule over the island. One month later, he told a Japanese parliamentary committee that "[Taiwan's] democracy is considerably matured and liberal economics is deeply ingrained, so it is a law-abiding country. In various ways, it is a country that shares a sense of values with Japan." At the same time, he admitted that "I know there will be a problem with calling [Taiwan] a country". Later, the Japanese Foreign Ministry tried to downplay or reinterpret his remarks.

In February 2007, the Royal Grenada Police Band played the National Anthem of the Republic of China in an inauguration of the reconstructed St George's Queen's Park Stadium funded by the PRC. Grenada had broken off diplomatic relationship with Taiwan just two years prior in favour of the PRC.

When the Kuomintang visited Mainland China in 2005, the government-controlled PRC media called this event a "visit," and called the KMT one of "Taiwan's political parties" even though the Kuomintang's full name remains the "Chinese

Nationalist Party." Interestingly in Mainland China, there is a legal party called the Revolutionary Committee of the Kuomintang that is officially one of the nine "consultative parties," according to the PRC's Chinese People's Political Consultative Conference.

## Possible Military Solutions and Intervention

Until 1979, both sides intended to resolve the conflict militarily. Intermittent clashes occurred throughout the 1950s and 1960s, with escalations comprising the First and Second Taiwan Strait crises. In 1979, with the U.S. change of diplomatic recognition to the PRC, the ROC lost its ally needed to "recover the mainland." Meanwhile, the PRC's desire to be accepted in the international community led it to promote peaceful unification under what would later be termed "one country, two systems", rather than to "liberate Taiwan" and to make Taiwan a Special Administrative Region.

### *PRC's Condition on Military Intervention*

Notwithstanding, the PRC government has issued triggers for an immediate war with Taiwan, most notably via its controversial Anti-Secession Law of 2005. These conditions are:

- if events occur leading to the *separation* of Taiwan from China *in any name*, or
- if a major event occurs which would lead to Taiwan's separation from China, or
- if all possibility of peaceful unification is lost.

It has been interpreted that these criteria encompass the scenario of Taiwan developing nuclear weapons.

Much saber-rattling by the PRC has been done over this, with Jiang Zemin, after assuming the mantle of the Chairman of the Central Military Commission, becoming a leading voice.

The third condition has especially caused a stir in Taiwan as the term "indefinitely" is open to interpretation. It has also been viewed by some as meaning that preserving the ambiguous status quo is not acceptable to the PRC, although the PRC stated on many occasions that there is no explicit timetable for reunification.

Concern over a formal declaration of *de jure* Taiwan independence is a strong impetus for the military buildup between Taiwan and mainland China. The former US Bush administration publicly declared that given the status quo, it would not aid Taiwan if it were to declare independence unilaterally.

According to President Chen Shuibian who was President of the Republic of China between 2000 and 2008, China accelerated the deployment of missiles against Taiwan up to 120 a year (May 2007), bringing the total arsenal to 706 ballistic missiles capable of being fitted with nuclear warheads that are aimed at Taiwan. Some believe that their deployment is a political tool on the part of the PRC to increase political pressure on Taiwan to abandon unilateral moves toward formal independence, at least for the time being, although the PRC government never declares such deployment publicly.. Legislative elections were held in Taiwan on January 12, 2008. The results gave the Kuomintang and the Pan-Blue Coalition a supermajority (86 of the 113 seats) in the legislature, handing a heavy defeat to President Chen Shuibian Democratic Progressive Party, which won the remaining 27 seats only. The junior partner in the Pan-Green Coalition, the Taiwan Solidarity Union, won no seats. The election for the 12th-term President and Vice-President of the Republic of China was held in the Republic of China (Taiwan) on Saturday, March 22, 2008. Kuomintang nominee Ma Ying-jeou won, with 58% of the vote, ending eight years of Democratic Progressive Party (DPP) presidential power. Along with the 2008 legislative election, Ma's landslide victory brought the Kuomintang back to power in Taiwan.

### *Balance of Power*

The possibility of war, the close geographical proximity of ROC-controlled Taiwan and PRC-controlled mainland China, and the resulting flare-ups that occur every few years, conspire to make this one of the most watched focal points in the Pacific. Both sides have chosen to have a strong naval presence. However, naval strategies between both powers greatly shifted in the 1980s and 1990s, while the ROC assumed a more defensive attitude by building and buying frigates and missile destroyers, and the PRC a more aggressive posture by developing long-range cruise missiles and supersonic surface-to-surface missiles.

Although the People's Liberation Army Air Force is considered large, most of its fleet consists of older generation J-7 fighters (localized MiG-21s and Mig-21BIs), raising doubts over the PLAAF's ability to control Taiwan's airspace in the event of a conflict. Since mid-1990s PRC has been purchasing, and later localizing, SU-27 based fighters. These Russian fighters, as well as their Chinese J11A variants, are currently over 170 in number, and have increased the effectiveness of PLAAF's Beyond Visual Range (BVR) capabilities. The introduction of 60 new-generation J10A fighters is anticipated to increase the PLAAF's firepower. PRC's acquisition of Russian Su 30 MKKs further enhanced the PLAAF's air-to-ground support ability. The ROC's air force, on the other hand, relies on Taiwan's second generation fighters, consisting of 150 US-built F-16 Fighting Falcons, approximately 60 French-built Mirage 2000-5s, and approximately 130 locally developed IDFs (Indigenous Defence Fighters). All of these ROC fighter jets are able to conduct BVR combat missions with BVR missiles, but the level of technology in mainland Chinese fighters is catching up. Also the United States Defence Intelligence Agency has reported that few of Taiwan's 400 total fighters are operationally capable.

In 2003, the ROC purchased four missile destroyers—the former USS *Kidd* class, and expressed a strong interest in the *Arleigh Burke* class. But with the growth of the PRC navy and air force, some doubt that the ROC could withstand a determined invasion attempt from mainland China in the future. These concerns have led to a view in certain quarters that Taiwanese independence, if it is to be implemented, should be attempted as early as possible, while the ROC still has the capacity to defend itself in an all-out military conflict. Over the past three decades, estimates of how long the ROC can withstand a full-scale invasion from across the Strait without any outside help have decreased from three months to only six days. Given such estimates, the US Navy has continued practicing "surging" its carrier groups, giving it the experience necessary to respond quickly to an attack on Taiwan. The US also collects data on the PRC's military deployments, through the use of spy satellites, for example. It would take days, if not weeks, for the PRC to prepare for a full assault on Taiwan.

However, numerous reports issued by the PRC, ROC and US militaries make mutually wild contradictory statements about the possible defence of Taiwan.

Naturally, war contingencies are not being planned in a vacuum. In 1979, the United States Congress passed the Taiwan Relations Act, a law generally interpreted as mandating U.S. defence of Taiwan in the event of an attack from the Chinese Mainland (the Act is applied to Taiwan and Penghu, but not to Jinmen or Matsu). The United States maintains the world's largest permanent fleet in the Pacific Region near Taiwan. The Seventh Fleet, operating primarily out of various bases in Japan, is a powerful naval contingent built upon the world's only permanently forward-deployed aircraft carrier USS George Washington (CVN-73). Although the stated purpose of the fleet is not Taiwanese defence, it can be safely assumed from past actions, that is one of the reasons why the fleet is stationed in those waters.

Starting in 2000, Japan renewed its defence obligations with the US and embarked on a rearmament program, partly in response to fears that Taiwan might be invaded. Some analysts believed that the PRC could launch preemptive strikes on military bases in Japan to deter US and Japanese forces from coming to the ROC's aid. Japanese strategic planners also see an independent Taiwan as vital, not only because the ROC controls valuable shipping routes, but also because its capture by PRC would make Japan more vulnerable. During World War II, the US invaded the Philippines, but another viable target to enable direct attacks on Japan would have been Taiwan (then known as Formosa). However, critics of the preemptive strike theory assert that the PRC would be loath to give Japan and the US such an excuse to intervene.

### *Third Taiwan Strait Crisis*

In 1996, the PRC began conducting military exercises near Taiwan, and launched several ballistic missiles over the island. The saber-rattling was done in response to the possible re-election of then President Lee Tenghui. The United States, under President Clinton, sent two aircraft carrier battle groups to the region, sailing them into the Taiwan Strait. The PRC, unable to track the ships' movements, and probably unwilling to escalate the conflict, quickly backed down. The event had little impact on the outcome of the election, since none of Lee's contestants were strong enough to defeat him, but it is widely believed that the PRC's aggressive acts, far from intimidating the Taiwanese population, gave Lee a boost that pushed his share of votes over 50 percent.

The possibility of war in the Taiwan Straits, even though quite low in the short-term, requires the PRC, ROC, and U.S. to remain wary and vigilant. The goal of the three parties at the moment seems to be, for the most part, to maintain the status quo.

## Developments Since 2004 and Future Prospects

### *Judicial*

On October 24, 2006, Dr. Roger C. S. Lin led a group of Taiwanese residents, including members of the Taiwan Nation Party, to file a Complaint for Declaratory Relief in the United States District Court for the District of Columbia. According to their lawyer, Mr. Charles Camp, "[t]he Complaint asks the Court to declare whether the Taiwanese plaintiffs, including members of the Taiwan Nation Party, have certain rights under the United States Constitution and other US laws". Their central argument is that, following Japanese renunciation of all rights and claims to Taiwan, Taiwan came under U.S. jurisdiction based on it being the principal occupying power as designated in the Treaty of Peace with Japan and remains so to this day. This case was opposed by the United States government.

The District Court agreed with United States government on March 18, 2008 and ruled that the case presents a political question; as such, the court concluded that it had no jurisdiction to hear the matter and dismissed the complaint. This decision has been appealed by plaintiffs and the appeals court unanimously upheld the district court ruling.

The PRC and Taiwan have agreed to increase cooperation in the area of law enforcement. Mainland police will begin staffing a liaison office in Taipei in 2010.

### Political

Although the situation is confusing, most observers believe that it is stable with enough understandings and gentlemen's agreements to keep things from breaking out into open warfare. The current controversy is over the term *one China*, as the PRC insists that the ROC must recognize this term to begin negotiations. Although the Democratic Progressive Party has moderated its support for Taiwan independence, there is still insufficient support within that party for former President Chen Shuibian to agree to one China. By contrast, the Kuomintang (KMT) and the People First Party (PFP) appear willing to agree to some variation of one China, and observers believed the position of the PRC was designed to sideline Chen until the 2004 presidential election where it was hoped that someone who was more supportive of Chinese reunification would come to power. Partly to counter this, Chen Shuibian announced in July 2002 that if the PRC does not respond to Taiwan's goodwill, Taiwan may "go on its own... road."

With Chen's re-election in 2004, Beijing's prospects for a speedier resolution were dampened, though they seemed strengthened again following the Pan-Blue majority in the 2004 legislative elections. However, public opinion in Taiwan reacted unfavorably towards the anti-secession law passed by the PRC in March 2005. Following two high profile visits by KMT and PFP party leaders to the PRC, the balance of public opinion appears to be ambiguous, with the Pan-Green Coalition gaining a majority in the 2005 National Assembly elections, but the Pan-Blue Coalition scoring a landslide victory in the 2005 municipal elections.

Legislative elections were held in Taiwan on January 12, 2008. The results gave the Kuomintang and the Pan-Blue Coalition an absolute majority (86 of the 113 seats) in the legislature, handing a heavy defeat to President Chen Shuibian Democratic Progressive Party, which won the remaining 27 seats. The junior partner in the Pan-Green Coalition, the Taiwan Solidarity Union, won no seats.

The election for the 12th President of ROC was held on March 22, 2008. Kuomintang candidate Ma Ying-jeou won, with 58% of the vote, ending eight years of Democratic

Progressive Party (DPP) leadership. Along with the 2008 legislative election, Ma's landslide victory brought the Kuomintang back to power in Taiwan. This new political situation has led to a decrease of tension between both sides of the Taiwan Strait and the increase of cross-strait relations, making a declaration of independence, or war, something unlikely.

Taiwan's Straits Exchange Foundation (SEF) and its Chinese counterpart – the Association for Relations Across the Taiwan Strait (ARATS) signed four agreements in Taipei on November 4, 2008. Both SEF and ARATS have agreed to address direct sea links, daily charter flights, direct postal service and food safety.

### *Public Opinion*

Public opinion in Taiwan regarding relations with the PRC is notoriously difficult to gauge, as poll results tend to be extremely sensitive to how the questions are phrased and what options are given, and there is a tendency by all political parties to spin the results to support their point of view.

According to a November 2005 poll from the Mainland Affairs Council, 37.7% of people living in the ROC favour maintaining the status quo until a decision can be made in the future, 18.4% favours maintaining the status quo indefinitely, 14% favours maintaining the status quo until eventual independence, 12% favours maintaining the status quo until eventual reunification, 10.3% favours independence as soon as possible, and 2.1% favours reunification as soon as possible. According to the same poll, 78.3% are opposed to the "One Country, Two Systems" model, which was used for Hong Kong and Macau, while 10.4% is in favour.

According to a June 2008 poll from a Taiwanese mainstream media TVBS, 58% of people living in Taiwan favour maintaining the status quo, 19% favours independence, and 8% favours unification. According to the same poll, if status quo is not an option and the ones who were surveyed must choose between "Independence" or "Unification", 65% are in favour of independence while 19% would opt for unification. The same poll also reveals that, in terms of self-identity, when the respondents are not told that a Taiwanese can also be a Chinese,

68% of the respondents identify themselves as "Taiwanese" while 18% would call themselves "Chinese". However, when the respondents are told that duo identity is an option, 45% of the respondents identify themselves as "Taiwanese only", 4% of the respondents call themselves "Chinese only" while 45% of the respondents call themselves "both Taiwanese as well as Chinese".

Furthermore, when it comes to preference in which national identity to be used in international organizations, 54% of people in the survey indicated that they prefer "Taiwan" and only 25% of the people voted for "Chinese Taipei".

According to an October 2008 poll from the Mainland Affairs Council, on the question of Taiwan's status, 36.17% of respondents favour maintaining the status quo until a decision can be made in the future, 25.53% favours maintaining the status quo indefinitely, 12.49% favours maintaining the status quo until eventual independence, 4.44% favours maintaining the status quo until eventual reunification, 14.80% favours independence as soon as possible, and 1.76% favours reunification as soon as possible. In the same poll, on the question of the PRC government's attitude towards the ROC government, 64.85% of the respondents consider the PRC government hostile or very hostile, 24.89 consider the PRC government friendly or very friendly, while 10.27% did not express an opinion. On the question of the PRC government's attitude towards the people in Taiwan, 45.98% of the respondents consider the PRC government hostile or very hostile, 39.6% consider the PRC government friendly or very friendly, while 14.43% did not express an opinion.

May 2009 Taiwan's (Republic of China) Department of the Interior published a survey examining whether people in Taiwan see themselves as Taiwanese, Chinese or both. 64.6% see themselves as Taiwanese, 11.5% as Chinese, 18.1% as both and 5.8% were unsure.

According to a December 2009 poll from a Taiwanese mainstream media TVBS, if status quo is not an option and the ones who were surveyed must choose between "Independence" or "Unification", 68% are in favour of independence while 13% would opt for unification.

### *Changing Taiwan's Status with Respect to the ROC Constitution*

From the perspective of the ROC constitution, which the mainstream political parties such as the KMT and DPP currently respect and recognize, (even though the procedures for the incorporation of Taiwan into ROC territory as specified in Article 4 have never been completed), changing the ROC's governing status or completely clarifying Taiwan's political status would at best require amending the ROC constitution. In other words, if reunification supporters wanted to reunity Taiwan with mainland China in such a way that would effectively abolish the ROC or affect the ROC's sovereignty, or if independence supporters wanted to abolish the ROC and establish a Republic of Taiwan, they would also need to amend or abolish the ROC constitution and redraft a new constitution. Passing an amendment requires an unusually broad political consensus, which includes approval from three-quarters of a quorum of members of the Legislative Yuan. This quorum requires at least three-quarters of all members of the Legislature. After passing the legislature, the amendments need ratification from at least fifty percent of all eligible voters of the ROC, *irrespective* of voter turnout.

Given these harsh constitutional requirements, neither the pan-greens nor pan-blues can unilaterally change Taiwan's political and legal status with respect to the ROC's constitution. However, extreme Taiwan independence supporters view the ROC's constitution as illegal and therefore believe that amendments to the ROC constitution are an invalid way to change Taiwan's political status.

## Note on Terminology

### *Political Status vs. Taiwan Issue*

Some scholarly sources as well as political entities like the PRC refer to Taiwan's controversial status as the "Taiwan question", "Taiwan issue", or "Taiwan problem". The ROC government does not like these terminologies, emphasizing that it should be called the "Mainland issue" or "Mainland question", because from the ROC's point of view, the PRC is making an issue out of or creating a problem out of Taiwan.

Others use the term "Taiwan Strait Question" because it implies nothing about sovereignty and because "Cross-Strait relations" is a term used by both the ROC and the PRC to describe their interactions. However, this term is also objectionable to some because it still implies that there is an issue, which they feel is created only by the PRC.

### *De-facto vs. De-jure*

The use of the terms *de-facto* and *de-jure* to describe Taiwan's as well as the Republic of China's status as a state is itself a contentious issue. This partially stems from the lack of precedents regarding derecognized, but still constitutionally functioning states. For instance, it is regularly argued that Taiwan satisfies the requirements of statehood at international law as stated in the Montevideo Convention. At the same time, there is continued debate on whether UN membership or recognition as a state by the UN is a decisive feature of statehood (since it represents broad recognition by the international community); the debate arises because non-state entities can often satisfy the Montevideo Convention factors, while the list of states recognised by the UN, for the most part, correlate well with entities recognised as states by customary international law. If the latter argument is accepted, then the Republic of China may have ceased to be a state post-1971 as a matter of international law ("*de jure*"), yet continued to otherwise function as the state that it previously was recognised as ("*de facto*").

From the 1990s onwards, media wire services sometimes describe Taiwan as having de-facto independence, whereas the Republic of China has always considered itself as a continuously functioning de-jure state.

## Economy

Taiwan's quick industrialization and rapid growth during the latter half of the twentieth century, has been called the "Taiwan Miracle" or "Taiwan Economic Miracle". As it has developed alongside Singapore, South Korea, and Hong Kong, Taiwan is one of the industrialized developed countries known as the "Four Asian Tigers".

Japanese rule prior to and during World War II brought forth changes in the public and private sectors of the economy, most notably in the area of public works, which enabled rapid communications and facilitated transport throughout much of the island. The Japanese also improved public education and made the system compulsory for all Taiwanese citizens during this time.

When the KMT government fled to Taiwan it brought the entire gold reserve and the foreign currency reserve of mainland China to the island which stabilized prices and reduced hyperinflation. More importantly, as part of its retreat to Taiwan, KMT brought with them the intellectual and business elites from mainland China. The KMT government instituted many laws and land reforms that it had never effectively enacted on mainland China. The government also implemented a policy of import-substitution, attempting to produce imported goods domestically. Much of this was made possible through US economic aid, subsidizing the higher cost of domestic production.

In 1962, Taiwan had a per capita gross national product (GNP) of $170, placing the island's economy squarely between Zaire and Congo. By 2008 Taiwan's per capita GNP, adjusted for purchasing power parity (PPP), had soared to $33,000 (2008 est.) contributing to a Human Development Index equivalent to that of other developed countries.

Today Taiwan has a dynamic capitalist, export-driven economy with gradually decreasing state involvement in investment and foreign trade. In keeping with this trend, some large government-owned banks and industrial firms are being privatized. Real annual growth in GDP has averaged about eight percent during the past three decades. Exports have provided the primary impetus for industrialization. The trade surplus is substantial, and foreign reserves are the world's fifth largest as of 31 December 2007.

Agriculture constitutes only two percent of the GDP, down from 35 percent in 1952. Traditional labour-intensive industries are steadily being moved offshore and with more capital and technology-intensive industries replacing them. Taiwan has become a major foreign investor in mainland China, Thailand, Indonesia, the Philippines, Malaysia, and Vietnam. It is estimated that some 50,000 Taiwanese businesses and 1,000,000 business people and their dependents are established in the PRC.

Because of its conservative financial approach and its entrepreneurial strengths, Taiwan suffered little compared with many of its neighbours from the 1997 Asian Financial Crisis. Unlike its ncighbours South Korea and Japan, the Taiwanese economy is dominated by small and medium sized businesses, rather than the large business groups. The global economic downturn, however, combined with poor policy coordination by the new administration and increasing bad debts in the banking system, pushed Taiwan into recession in 2001, the first whole year of negative growth since 1947. Due to the relocation of many manufacturing and labour intensive industries to mainland China, unemployment also reached a level not seen since the 1973 oil crisis. This became a major issue in the 2004 presidential election. Growth averaged more than 4% in the 2002–2006 period and the unemployment rate fell below 4%. Since the global financial crisis starting with United States in 2007, the unemployment rate has risen to over 5.9% and Economic Growth fallen to-2.9%.

Leading technologies of Taiwan include:

- Bicycle manufacturing, ex: Giant Bicycles, Merida
- Biotechnology
- Semiconductor device fabrication, TSMC
- Laptops, ex: Acer, Asus, BenQ
- Smart phones, ex: HTC
- Communication & Network, ex: D-Link
- Technology, ex: Garmin
- Shipping, ex: Evergreen.

# 5

# Japan and Island Kingdom

Japan is an island country in East Asia. Located in the Pacific Ocean, it lies to the east of the Sea of Japan, China, North Korea, South Korea and Russia, stretching from the Sea of Okhotsk in the north to the East China Sea and Taiwan in the south. The characters that make up Japan's name mean "sun-origin", which is why Japan is sometimes referred to as the "Land of the Rising Sun".

Japan is an archipelago of 6,852 islands. The four largest islands are Honshu, Hokkaido, Kyushu and Shikoku, together accounting for 97% of Japan's land area. Most of the islands are mountainous, many volcanic; for example, Japan's highest peak, Mount Fuji, is a volcano. Japan has the world's tenth-largest population, with over 128 million people. The Greater Tokyo Area, which includes the *de facto* capital city of Tokyo and several surrounding prefectures, is the largest metropolitan area in the world, with over 30 million residents.

Archaeological research indicates that people were living on the islands of Japan as early as the Upper Paleolithic period. The first written mention of Japan begins with brief appearances in Chinese history texts from the first century A.D. Influence from the outside world followed by long periods of isolation has characterized Japan's history. Since adopting its constitution in 1947, Japan has maintained a unitary constitutional monarchy with an emperor and an elected parliament called the Diet.

A major economic power, Japan has the world's second-largest economy by nominal GDP and the third largest in purchasing power parity. It is also the world's fourth largest exporter and fifth largest importer. It is also the only Asian country in the G8 and is currently serving as a non-permanent member of the UN Security Council.

Although Japan has officially renounced its right to declare war, it maintains a modern and extensive military force which is employed in self-defence and peacekeeping roles. It is a developed country with very high living standards (10th highest HDI). Japan has the highest life expectancy of any country in the world (according to both the UN and WHO estimates) and the third lowest infant mortality rate.

## Etymology

The English word *Japan* is an exonym. The Japanese names for Japan are *Nippon*, and *Nihon*. They are both written in Japanese using the kanji. The Japanese name *Nippon* is used for most official purposes, including on Japanese money, postage stamps, and for many international sporting events. *Nihon* is a more casual term and the most frequently used in contemporary speech. Japanese people refer to themselves as *Nihonjin*) and they call their language *Nihongo*.

Both *Nippon* and *Nihon* literally mean "the sun's origin" and are often translated as the *Land of the Rising Sun*. This nomenclature comes from Imperial correspondence with the Chinese Sui Dynasty and refers to Japan's eastward position relative to China. Before *Nihon* came into official use, Japan was known as *Wa* or *Wakoku*.

The English word for Japan came to the West from early trade routes. The early Mandarin or possibly Wu Chinese word for Japan was recorded by Marco Polo as *Cipangu*. In modern Shanghainese, a Wu dialect, the pronunciation of characters 'Japan' is *Zeppen*; in Wu, the character has two pronunciations, informal and formal. (In some southern Wu dialects, is pronounced, similar to its pronunciation in Japanese.)

The old Malay word for Japan, *Jepang* (now spelled *Jepun* in Malaysia, though still spelled *Jepang* in Indonesia), was borrowed from a Chinese language, and this Malay word was encountered by Portuguese traders in Malacca in the 16th century. It is thought the Portuguese traders were the first to bring the word to Europe. It was first recorded in English in a 1565 letter spelled *Giapan*.

## History

The first signs of occupation on the Japanese Archipelago appeared witn a Paleolithic culture around 30,000 BC, followed from around 14,000 BC by the Jomon period, a Mesolithic to Neolithic semi-sedentary hunter-gatherer (possibly Ainu) culture of pit dwelling and a rudimentary form of agriculture. Decorated clay vessels from this period, often with plaited patterns, are some of the oldest surviving examples of pottery in the world.

The Yayoi period, starting around 500 BC, saw the introduction of many new practices, such as wet-rice farming, a new style of pottery and Metallurgy brought by migrants from China and Korea.

The Japanese first appear in written history in China's *Book of Han*. According to the Chinese *Records of Three Kingdoms*, the most powerful kingdom on the archipelago during the third century was called Yamataikoku.

Buddhism was first introduced to Japan from Baekje, one of the Three Kingdoms of Korea, but the subsequent development of Japanese Buddhism and Buddhist sculptures were primarily influenced by China. Despite early resistance, Buddhism was promoted by the ruling class and eventually gained growing acceptance since the Asuka period.

The Nara period of the eighth century marked the first emergence of a strong central Japanese state, centered on an imperial court in the city of Heijo-kyo, or modern-day Nara. In addition to the continuing adoption of Chinese administrative practices, the Nara period is characterized by the appearance of a nascent written literature with the completion of the massive chronicles *Kojiki* (712) and *Nihon Shoki* (720). (Nara was not the first capital city in Japan, though. Before Nara, Fujiwara-kyo and Asuka served as capitals of the Yamato state.) The smallpox epidemic of 735-737 is believed to had killed as many as one-third of Japan's population.

In 784, Emperor Kammu moved the capital from Nara to Nagaoka-kyo for a brief ten-year period, before relocating it to

Heian-kyo (modern-day Kyoto) in 794, where it remained for more than a millennium. This marked the beginning of the Heian period, during which time a distinctly indigenous Japanese culture emerged, noted for its art, poetry and literature. Lady Murasaki's *The Tale of Genji* and the lyrics of modern Japan's national anthem, *Kimi ga Yo* were written during this time.

Japan's feudal era was characterized by the emergence of a ruling class of warriors, the samurai. In 1185, following the defeat of the rival Taira clan, Minamoto no Yoritomo was appointed Shogun and established a base of power in Kamakura. After Yoritomo's death, the Hojo clan came to rule as regents for the shoguns. Zen Buddhism was introduced from China in the Kamakura period (1185–1333) and became popular among the samurai class.

The Kamakura shogunate managed to repel Mongol invasions in 1274 and 1281, aided by a storm that the Japanese interpreted as a *kamikaze*, or Divine Wind. The Kamakura shogunate was eventually overthrown by Emperor Go-Daigo, who was soon himself defeated by Ashikaga Takauji in 1336. The succeeding Ashikaga shogunate failed to control the feudal warlords (*daimyo*), and a civil war erupted (the Onin War) in 1467 which opened a century-long Sengoku ("Warring States") period.

During the sixteenth century, traders and Jesuit missionaries from Portugal reached Japan for the first time, initiating active commercial and cultural exchange between Japan and the West (*Nanban trade*).

Oda Nobunaga conquered numerous other daimyo by using European technology and firearms and had almost unified the nation when he was assassinated in 1582. Toyotomi Hideyoshi succeeded Nobunaga and united the nation in 1590. Hideyoshi invaded Korea twice, but following several defeats by Korean and Ming China forces and Hideyoshi's death, Japanese troops were withdrawn in 1598.

After Hideyoshi's death, Tokugawa Ieyasu utilized his position as regent for Hideyoshi's son Toyotomi Hideyori to gain political and military support. When open war broke out, he defeated rival clans in the Battle of Sekigahara in 1600.

Ieyasu was appointed shogun in 1603 and established the Tokugawa shogunate at Edo (modern Tokyo). The Tokugawa shogunate enacted a variety of measures such as *Buke shohatto* to control the autonomous daimyo.

In 1639, the shogunate began the isolationist *sakoku* ("closed country") policy that spanned the two and a half centuries of tenuous political unity known as the Edo period. The study of Western sciences, known as *rangaku*, continued during this period through contacts with the Dutch enclave at Dejima in Nagasaki. The Edo period also gave rise to *kokugaku*, or literally "national studies", the study of Japan by the Japanese themselves. According to one authority, there were at least 130 famines during the Edo period, of which 21 were particularly serious.

On March 31, 1854, Commodore Matthew Perry and the "Black Ships" of the United States Navy forced the opening of Japan to the outside world with the Convention of Kanagawa. Subsequent similar treaties with the Western countries in the Bakumatsu period brought Japan into economic and political crises. The abundance of the prerogative and the resignation of the shogunate led to the Boshin War and the establishment of a centralized state unified under the name of the Emperor (Meiji Restoration).

Adopting Western political, judicial and military institutions, the Cabinet organized the Privy Council, introduced the Meiji Constitution, and assembled the Imperial Diet. The Meiji Restoration transformed the Empire of Japan into an industrialized world power that embarked on a number of military conflicts to expand the nation's sphere of influence. After victories in the First Sino-Japanese War (1894–1895) and the Russo-Japanese War (1904–1905), Japan gained control of Taiwan, Korea, and the southern half of Sakhalin. Japan's population increased from 35 million in 1873 to 70 million in 1935.

## *20th Century*

The early twentieth century saw a brief period of "Taisho democracy" overshadowed by the rise of expansionism and militarization. World War I enabled Japan, which joined the side of the victorious Allies, to expand its influence and territorial holdings. Japan continued its expansionist policy by occupying Manchuria in 1931. As a result of international condemnation for this occupation, Japan resigned from the League of Nations two years later. In 1935, local assemblies were established in Taiwan. In 1936, Japan signed the Anti-Comintern Pact with Nazi Germany, joining the Axis powers in 1941. In 1941, Japan signed the Soviet–Japanese Neutrality Pact with Soviet Union, respecting both Manchukou and Mongolian People's Republic territories.

In 1937, the Empire of Japan invaded other parts of China, precipitating the Second Sino-Japanese War (1937–1945). In 1940, the Empire then invaded French Indochina, after which the United States placed an oil embargo on Japan. On December 7, 1941, Japan attacked the United States naval base in Pearl Harbor and declared war on the United States, the United Kingdom and Netherlands.

This act brought the United States into World War II and, on December 8, these three countries declared war on Japan. After the atomic bombings of Hiroshima and Nagasaki in 1945, along with the Soviet Union joining the war against it, Japan agreed to an unconditional surrender of all Japanese forces on August 15 (Victory over Japan Day).

The war cost Japan and countries part of the Greater East Asia Co-Prosperity Sphere millions of lives and left much of the country's industry and infrastructure destroyed. The Allied powers repatriated millions of ethnic Japanese from colonies throughout Asia. The International Military Tribunal for the Far East, was convened by the Allies (on May 3, 1946) to prosecute some Japanese leaders for war crimes. However, all members of the bacteriological research units and members of the imperial family involved in the conduct of the war were exonerated from criminal prosecutions by the Supreme Commander of the Allied Forces.

In 1947, Japan adopted a new pacifist constitution emphasizing liberal democratic practices. The Allied occupation ended by the Treaty of San Francisco in 1952 and Japan was granted membership in the United Nations in 1956. Japan later achieved spectacular growth to become the second largest economy in the world, with an annual growth rate averaging 10% for four decades. This ended in the mid-1990s when Japan suffered a major recession. Positive growth in the early twenty-first century has signaled a gradual recovery.

## Government and Politics

Japan is a constitutional monarchy where the power of the Emperor is very limited. As a ceremonial figurehead, he is defined by the constitution as "the symbol of the state and of the unity of the people". Power is held chiefly by the Prime Minister of Japan and other elected members of the Diet, while sovereignty is vested in the Japanese people. The Emperor effectively acts as the head of state on diplomatic occasions. Akihito is the current Emperor of Japan. Naruhito, Crown Prince of Japan, stands as next in line to the throne.

Japan's legislative organ is the National Diet, a bicameral parliament. The Diet consists of a House of Representatives, containing 480 seats, elected by popular vote every four years or when dissolved and a House of Councillors of 242 seats, whose popularly elected members serve six-year terms. There is universal suffrage for adults over 20 years of age, with a secret ballot for all elective offices. In 2009, the social liberal Democratic Party of Japan took power after 54 years of the liberal conservative Liberal Democratic Party's rule.

The Prime Minister of Japan is the head of government. The position is appointed by the Emperor of Japan after being designated by the Diet from among its members and must enjoy the confidence of the House of Representatives to remain in office. The Prime Minister is the head of the Cabinet (the literal translation of his Japanese title is "Prime Minister of the Cabinet") and appoints and dismisses the Ministers of State, a majority of whom must be Diet members. Naoto Kan has been designated by the Diet to replace Yukio Hatoyama as the Prime Minister of Japan. He is awaiting confirmation from the Emperor before he will officially be Prime Minister.

Historically influenced by Chinese law, the Japanese legal system developed independently during the Edo period through texts such as *Kujikata Osadamegaki*. However, since the late nineteenth century, the judicial system has been largely based on the civil law of Europe, notably France and Germany. For example, in 1896, the Japanese government established a civil code based on the German model. With post-World War II modifications, the code remains in effect in present-day Japan. Statutory law originates in Japan's legislature, the National Diet of Japan, with the rubber stamp approval of the Emperor. The current constitution requires that the Emperor promulgates legislation passed by the Diet, without specifically giving him the power to oppose the passing of the legislation. Japan's court system is divided into four basic tiers: the Supreme Court and three levels of lower courts. The main body of Japanese statutory law is a collection called the Six Codes.

## *Foreign Relations and Military*

Japan maintains close economic and military relations with its key ally the United States, with the U.S.-Japan security alliance serving as the cornerstone of its foreign policy. A member state of the United Nations since 1956, Japan has served as a non-permanent Security Council member for a total of 19 years, most recently for 2009 and 2010. It is also one of the G4 nations seeking permanent membership in the Security Council.

As a member of the G8, the APEC, the "ASEAN Plus Three" and a participant in the East Asia Summit, Japan actively participates in international affairs and enhances diplomatic ties with its important partners around the world. Japan signed a security pact with Australia in March 2007 and with India in October 2008. It is also the world's third largest donor of official development assistance after the United States and United Kingdom, donating US$8.86 billion in 2004. Japan contributed non-combatant troops to the Iraq War but subsequently withdrew its forces from Iraq. The Japanese Maritime Self Defence Force is a regular participant in RIMPAC maritime exercises.

Japan is engaged in several territorial disputes with its neighbours: with Russia over the South Kuril Islands, with South Korea over the Liancourt Rocks, with the People's Republic of China and Republic of China (Taiwan) over the Senkaku Islands, and with the PRC over the EEZ around Okinotorishima.

Japan also faces an ongoing dispute with North Korea over its abduction of Japanese citizens and its nuclear weapons and missile program. As a result of the Kuril Islands dispute, Japan is technically still at war with Russia since no treaty resolving the issue was ever signed.

Japan's military is restricted by the Article 9 of the Japanese Constitution, which renounces Japan's right to declare war or use military force as a means of settling international disputes. Japan's military is governed by the Ministry of Defence, and primarily consists of the Japan Ground Self-Defence Force (JGSDF), the Japan Maritime Self-Defence Force (JMSDF) and the Japan Air Self-Defence Force (JASDF). The forces have been recently used in peacekeeping operations and the deployment of Japanese troops to Iraq marked the first overseas use of its military since World War II.

### *Administrative Divisions*

Japan consists of forty-seven prefectures, each overseen by an elected governor, legislature and administrative bureaucracy. Each prefecture is further divided into cities, towns and villages.

The nation is currently undergoing administrative reorganization by merging many of the cities, towns and villages with each other. This process will reduce the number of sub-prefecture administrative regions and is expected to cut administrative costs.

Japan has dozens of major cities, which play an important role in Japan's culture, heritage and economy.

### *Geography*

Japan is an island nation in East Asia comprising a stratovolcanic archipelago extending along the Pacific coast of Asia. Measured from the geographic coordinate system, Japan is 36° north of the equator and 138° east of the Prime Meridian. The country is east-notheast of China and Taiwan (separated by the East China Sea) and slightly east of Korea (separated by the Sea of Japan). The country is south of the Russian Far East.

The major islands, sometimes called the "Home Islands", are (from north to south) Hokkaido, Honshu (the "mainland"), Shikoku and Kyushu. There are also about 3,000 smaller islands, including Okinawa, and islets, some inhabited and others uninhabited. In total, as of 2006, Japan's territory is 377,923.1 km$^2$, of which 374,834 km$^2$ is land and 3,091 km$^2$

water. This makes Japan's total area slightly smaller than the U.S. state of Montana, slightly bigger than Germany.

### *Statistics*

Location: Eastern Asia, island chain between the North Pacific Ocean and the Sea of Japan, east of the Korean Peninsula.

Area:

- *total*: 377,835 km$^2$
- *land*: 374,744 km$^2$
- *water*: 3,091 km$^2$

*Notes*: Includes the Bonin Islands, Daito-shoto, Marcus Island, Okino-tori-shima, the Ryukyu Islands, and the Volcano Islands. Ownership of the Liancourt Rocks (Japanese: Takeshima, Korean: *Dokdo*) is in dispute.

***Area comparative:*** slightly smaller than California, USA

***Land boundaries:*** none

***Coastline:*** 34,751 km (21,593 mi).

Maritime claims:

- *Exclusive economic zone*: 200 nmi (370.4 km; 230.2 mi)
- *Territorial sea*: 12 nmi (22.2 km; 13.8 mi); between 3 and 12 nmi (5.6 and 22.2 km; 3.5 and 13.8 mi) in the international straits—La Perouse or Soya Strait, Tsugaru Strait, Osumi, and Eastern and Western Channels of the Korea or Tsushima Strait.

***Climate:*** varies from tropical in south to cool temperate in north.

***Terrain:*** mostly rugged and mountainous.

***Natural resources:*** negligible coal, oil, iron resources, fish, and mineral resources.

Land use:

- arable land: 11%
- permanent crops: 1%
- permanent pastures: 2%
- forests and woodland: 67%
- other: 19% (1993 est.)

Irrigated land: 27,820 km$^2$ (1993 est.) 73% of Japan is mountains.

## *Composition and Topography*

About 73% of Japan is mountainous, with a mountain range running through each of the main islands. Japan's highest mountain is Mt. Fuji, with an elevation of 3,776 m (12,388 ft). Since so very little flat area exists, many hills and mountainsides are cultivated all the way to the top. As Japan is situated in a volcanic zone along the Pacific deeps, frequent low-intensity earth tremors and occasional volcanic activity are felt throughout the islands. Destructive earthquakes occur several times a century. Hot springs are numerous and have been exploited as an economic capital by the leisure industry.

The mountainous islands of the Japanese Archipelago form a crescent off the eastern coast of Asia. They are separated from the mainland by the Sea of Japan, which historically served as a protective barrier. The country consists of four principal islands: Hokkaido, Honshu, Shikoku, and Kyushu; more than 3,000 adjacent islands and islets, including Izu Oshima in the Nanpo Islands; and more than 200 other smaller islands, including those of the Amami, Okinawa, and Sakishima chains of the Ryukyu Islands. The national territory also includes the small Bonin or Ogasawara Islands, which include Iwo Jima and the Volcano Islands (Kazan Retto), stretching some 1,100 kilometers from the main islands. A territorial dispute with Russia, dating from the end of World War II, over the two southernmost of the Kuril Islands, Etorofu (Iturup) and Kunashiri, and the smaller Shikotan and Habomai Islands northeast of Hokkaido remains a sensitive spot in Japanese-Russian relations as of 2005. Excluding disputed territory, the archipelago covers about 377,000 square kilometers. No point in Japan is more than 150 kilometers from the sea.

The four major islands are separated by narrow straits and form a natural entity. The Ryukyu Islands curve 970 kilometers southward from Kyushu.

The distance between Japan and the Korean Peninsula, the nearest point on the Asian continent, is about 200 kilometers at the Korea Strait. Japan has always been linked with the continent through trade routes, stretching in the north toward

Siberia, in the west through the Tsushima Islands to the Korean Peninsula, and in the south to the ports on the south China coast.

The Japanese islands are the summits of mountain ridges uplifted near the outer edge of the continental shelf. About 73 percent of Japan's area is mountainous, and scattered plains and intermontane basins (in which the population is concentrated) cover only about 25 percent. A long chain of mountains runs down the middle of the archipelago, dividing it into two halves, the "face," fronting on the Pacific Ocean, and the "back," toward the Sea of Japan. On the Pacific side are steep mountains 1,500 to 3,000 meters high, with deep valleys and gorges. Central Japan is marked by the convergence of the three mountain chains—the Hida, Kiso, and Akaishi mountains—that form the Japanese Alps (Nihon Arupusu), several of whose peaks are higher than 3,000 meters. The highest point in the Japanese Alps is Mount Kita at 3,193 meters. The highest point in the country is Mount Fuji (Fujisan, also erroneously called Fujiyama), a volcano dormant since 1707 that rises to 3,776 meters above sea level in Shizuoka Prefecture. On the Sea of Japan side are plateaus and low mountain districts, with altitudes of 500 to 1,500 meters.

None of the populated plains or mountain basins are extensive in area. The largest, the Kanto Plain, where Tokyo is situated, covers only 13,000 square kilometers. Other important plains are the Nobi Plain surrounding Nagoya, the Kinki Plain in the Osaka-Kyoto area, the Sendai Plain around the city of Sendai in northeastern Honshu, and the Ishikari Plain on Hokkaido. Many of these plains are along the coast, and their areas have been increased by reclamation throughout recorded history.

The small amount of habitable land has prompted significant human modification of the terrain over many centuries. Land was reclaimed from the sea and from river deltas by building dikes and drainage, and rice paddies were built on terraces carved into mountainsides. The process continued in the modern period with extension of shorelines and building of artificial islands for industrial and port development, such as Port Island in Kobe and the new Kansai International Airport in Osaka Bay. Hills and even mountains have been razed to provide flat areas for housing. Rivers are generally steep and swift, and few are suitable for navigation except in their lower reaches. Most rivers are fewer than 300 kilometers in length, but their rapid flow from the mountains provides a valuable, renewable resource: hydroelectric power generation. Japan's hydroelectric power potential has been exploited almost to capacity. Seasonal variations in flow have led to extensive development of flood control measures. Most of the rivers are very short. The longest, the Shinano River, which winds through Nagano Prefecture to Niigata Prefecture and flows into the Sea of Japan, is only 367 kilometers long. The largest freshwater lake is Lake Biwa, northeast of Kyoto.

Extensive coastal shipping, especially around the Inland Sea (Seto Naikai), compensates for the lack of navigable rivers. The Pacific coastline south of Tokyo is characterized by long, narrow, gradually shallowing inlets produced by sedimentation, which has created many natural harbours. The Pacific coastline north of Tokyo, the coast of Hokkaido, and the Sea of Japan coast are generally unindented, with few natural harbours.

### *Climate*

Japan belongs to the temperate zone with four distinct seasons, but its climate varies from cool temperate in the north to subtropical in the south. The climate is also affected by the seasonal winds that blow from the continent to the ocean in winters and vice versa in summers.

Japan is generally a rainy country with high humidity. Because of its wide range of latitude, Japan has a variety of climates, with a range often compared to that of the east coast of North America, from Nova Scotia to Georgia. Tokyo is at about 35 degrees north latitude, comparable to that of Tehran, Athens, or Las Vegas. The generally humid, temperate climate exhibits marked seasonal variation celebrated in art and literature, as well as regional variations ranging from cool in Hokkaido to subtropical in Kyushu. Climate also varies with altitude and with location on the Pacific Ocean or on the Sea of Japan. Northern Japan has warm summers but long, cold winters with heavy snow. Central Japan has hot, humid summers and short winters, and southwestern Japan has long, hot, humid summers and mild winters.

Two primary factors influence Japan's climate: a location near the Asian continent and the existence of major oceanic currents. The climate from June to September is marked by hot, wet weather brought by tropical airflows from the Pacific Ocean and Southeast Asia. These airflows are full of moisture and deposit substantial amounts of rain when they reach land. There is a marked rainy season, beginning in early June and continuing for about a month. It is followed by hot, sticky weather. Five or six typhoons pass over or near Japan every year from early August to early September, sometimes resulting in significant damage. Annual precipitation, which averages between 1,000 and 2,000 mm (39.4 and 78.7 in), is concentrated in the period between June and September. In fact, 70 to 80 percent of the annual precipitation falls during this period. In winter, a high-pressure area develops over Siberia, and a low-pressure area develops over the northern Pacific Ocean. The result is a flow of cold air eastward across Japan that brings freezing temperatures and heavy snowfalls to the central mountain ranges facing the Sea of Japan, but clear skies to areas fronting on the Pacific.

Two major ocean currents affect this climatic pattern: the warm Kuroshio Current (Black Current; also known as the Japan Current); and the cold Oyashio Current (Parent Current; also known as the Okhotsk Current). The Kuroshio Current flows northward on the Pacific side of Japan and warms areas as far north as Tokyo; a small branch, the Tsushima Current, flows up the Sea of Japan side. The Oyashio Current, which abounds in plankton beneficial to coldwater fish, flows southward along the northern Pacific, cooling adjacent coastal areas. The meeting point of these currents at 36 north latitude is a bountiful fishing ground.

Late June and early July are a rainy season—except in Hokkaido—as a seasonal rain front or *baiu zensen* stays above Japan. In summer and early autumn, typhoons, grown from tropical depressions generated near the equator, attack Japan with furious rainstorms. Its varied geographical features divide Japan into six principal climatic zones.

- Hokkaido: Belonging to the cool temperate zone, Hokkaido has long, cold winters and cool summers. Precipitation is not great.

- Sea of Japan: The northwest seasonal wind in winter gives heavy snowfalls. In summer it is less hot than in the Pacific area but sometimes experiences extreme high temperatures due to the foehn wind phenomenon.
- Central Highland: A typical inland climate gives large temperature differences between summers and winters and between days and nights. Precipitation is not large throughout the year.
- Seto Inland Sea: The mountains in the Chugoku and Shikoku regions block the seasonal winds and bring mild climate and many fine days throughout the year.
- Pacific Ocean: Winters are cold, with little snowfall, and summers are hot and humid due to the southeast seasonal wind.
- Southwest Islands: This zone has a subtropical climate with warm winters and hot summers. Precipitation is very high, and is especially affected by the rainy season and typhoons.

The warmest winter temperatures are found in the Nanpo and Bonin Islands, which enjoy a tropical climate due to the combination of latitude, distance from the Asian mainland, and warming effect of winds from the Kuroshio, as well as the Volcano Islands.

As an island nation, Japan has a long coastline. A few prefectures are landlocked: Gunma, Tochigi, Saitama, Nagano, Yamanashi, Gifu, Shiga, and Nara. As Mt. Fuji provides rain shadow effects in Yamanashi it has the least rainfall in Japan, which still exceeds 1,000 mm annually. The others all have coasts on the Pacific Ocean, Sea of Japan, Seto Inland Sea or have a body of salt water connected to them. Two prefectures—Hokkaido and Okinawa—are composed entirely of islands.

The hottest temperature ever measured in Japan, 40.9 °C (105.6 °F), occurred in Tajimi, Gifu on August 16, 2007.

## Environmental Protection

Environment-current issues: In the 2006 environment annual report, the Ministry of Environment reported that current major issues are global warming and preservation of the ozone layer, conservation of the atmospheric environment, water and soil, waste management and recycling, measures for chemical substances, conservation of the natural environment and the participation in the international cooperation.

### *Environment-International Agreements*

***Party to:*** Antarctic-Environmental Protocol, Antarctic Treaty, Biodiversity, Climate Change, Desertification, Endangered Species, Environmental Modification, Hazardous Wastes (Basel Convention), Law of the Sea, Marine Dumping, Nuclear Test Ban, Ozone Layer Protection (Montreal Protocol), Ship Pollution (MARPOL 73/78), Tropical Timber 83, Tropical Timber 94, Wetlands (Ramsar Convention), Whaling.

***Signed and ratified:*** Climate Change-Kyoto Protocol.

### *Natural Hazards*

Ten percent of the world's active volcanoes — forty in the early 1990s (another 148 were dormant) — are found in Japan, which lies in a zone of extreme crustal instability. As many as 1,500 earthquakes are recorded yearly, and magnitudes of four to six on the Richter scale are common. Minor tremors occur almost daily in one part of the country or another, causing slight shaking of buildings. Major earthquakes occur infrequently; the most famous in the twentieth century was the great Kanto earthquake of 1923, in which 130,000 people died. Undersea earthquakes also expose the Japanese coastline to danger from tsunamis and tidal waves.

Japan has become a world leader in research on causes and prediction of earthquakes. The development of advanced technology has permitted the construction of skyscrapers even in earthquake-prone areas. Extensive civil defence efforts focus on training in protection against earthquakes, in particular against accompanying fire, which represents the greatest danger.

Another common hazard are typhoons that reach Japan from the Pacific.

### *Regions*

Japan is informally divided into eight regions. Each contains several prefectures, except the Hokkaido region, which covers only Hokkaido Prefecture.

The region is not an official administrative unit, but has been traditionally used as the regional division of Japan in a number of contexts: for example, maps and geography textbooks divide Japan into the eight regions, weather reports usually give the weather by region, and many businesses and institutions use their home region as part of their name (Kinki Nippon Railway, Chugoku Bank, Tohoku University, etc.). While Japan has eight High Courts, their jurisdictions do not correspond to the eight regions..

### *Extreme Points*

This is a list of the extreme points of Japan, the points that are farther north, south, east or west than any other location.

#### *Japan*

- Northernmost point
  - o Cape Soya, Wakkanai, Hokkaido – *45°31'N*, 141°56'E
  - o Including land currently disputed with Russia: Cape Kamuiwakka, Iturup – *45°33'N*, 148°45'E
- Southernmost point: Okino Torishima – *20°25'N*, 136°04'E
- Westernmost point: Yonaguni – 24°272 N, *122°592 E*
- Easternmost point: Minami Torishima – 24°182 N, *153°582 E.*

#### *Japan (Main Islands)*

- Northernmost point: Cape Soya, Wakkanai, Hokkaido – *45°31'N*, 141°56'E
- Southernmost point: Cape Sata on Osumi Peninsula, Minamiosumi, Kagoshima – *30°59'N*, 130°39'E
- Westernmost point: Kosakibana, Sasebo (formerly Kosaza), Nagasaki – 33°13'N, *129°33'E*
- Easternmost point: Cape Nosappu, Nemuro, Hokkaido – 43°22'N, *145°492 E.*

### *Elevation Extremes*

- Lowest point: Hachiro-gata –-4 m
- Highest point: Mount Fuji – 3,776 m.

Japan is a country of over three thousand islands extending along the Pacific coast of Asia. The main islands, running from north to south, are Hokkaido, Honshu (the main island), Shikoku and Kyushu. The Ryukyu Islands, including Okinawa, are a chain of islands south of Kyushu. Together they are often known as the Japanese Archipelago.

About 70% to 80% of the country is forested, mountainous, and unsuitable for agricultural, industrial, or residential use. This is because of the generally steep elevations, climate and risk of landslides caused by earthquakes, soft ground and heavy rain. This has resulted in an extremely high population density in the habitable zones that are mainly located in coastal areas. Japan is one of the most densely populated countries in the world.

Its location on the Pacific Ring of Fire, at the juncture of three tectonic plates, gives Japan frequent low-intensity tremors and occasional volcanic activity. Destructive earthquakes, often resulting in tsunamis, occur several times each century. The 1923 Tokyo earthquake killed over 140,000. The most recent major quakes are the 2004 Chuetsu earthquake and the Great Hanshin Earthquake of 1995. Hot springs are numerous and have been developed as resorts.

## Geology

The islands of Japan are primarily the result of several large oceanic movements occurring over hundreds of millions of years from the mid-Silurian to the Pleistocene as a result of the subduction of the Philippine Sea Plate beneath the continental Amurian Plate and Okinawa Plate to the south, and subduction of the Pacific Plate under the Okhotsk Plate to the north.

Japan was originally attached to eastern coast of the Eurasian continent. The subducting plates, being deeper than the Eurasian plate, pulled Japan eastward, opening the Sea of Japan around 15 million years ago. The Strait of Tartary and the Korea Strait opened much later.

Japan is situated in a volcanic zone on the Pacific Ring of Fire. Frequent low intensity earth tremors and occasional volcanic activity are felt throughout the islands. Destructive earthquakes, often resulting in tsunamis, occur several times a century. The most recent major quakes include the 2004 Chuetsu earthquake and the Great Hanshin Earthquake of 1995. Hot springs are numerous and have been developed as resorts.

## *Climate*

The climate of Japan is predominantly temperate, but varies greatly from north to south. Japan's geographical features divide it into six principal climatic zones:

- Hokkaido: The northernmost zone has a temperate climate with long, cold winters and cool summers. Precipitation is not heavy, but the islands usually develop deep snow banks in the winter.
- Sea of Japan: On Honshu's west coast, the northwest wind in the wintertime brings heavy snowfall. In the summer, the region is cooler than the Pacific area, though it sometimes experiences extremely hot temperatures, because of the foehn wind phenomenon.
- Central Highland: A typical inland climate, with large temperature differences between summer and winter, and between day and night. Precipitation is light.
- Seto Inland Sea: The mountains of the Chugoku and Shikoku regions shelter the region from the seasonal winds, bringing mild weather throughout the year.
- Pacific Ocean: The east coast experiences cold winters with little snowfall and hot, humid summers because of the southeast seasonal wind.
- Ryukyu Islands: The Ryukyu Islands have a subtropical climate, with warm winters and hot summers. Precipitation is very heavy, especially during the rainy season. Typhoons are common.

The highest temperature ever measured in Japan—40.9 °C (105.6 °F)—was recorded on August 16, 2007. The main rainy season begins in early May in Okinawa, and the stationary rain front responsible for this gradually works its way north until it dissipates in northern Japan before reaching Hokkaido in late July. In most of Honshu, the rainy season begins before the middle of June and lasts about six weeks. In late summer and early autumn, typhoons often bring heavy rain.

## *Ecology*

Japan has nine forest ecoregions which reflect the climate and geography of the islands. They range from subtropical moist broadleaf forests in the Ryukyu and Bonin islands, to temperate broadleaf and mixed forests in the mild climate regions of the main islands, to temperate coniferous forests in the cold, winter portions of the northern islands.

## *Environment*

Japan's environmental history and current policies reflect a balance between economic development and environmental protection. In the rapid economic growth after World War II, environmental policies were downplayed by the government and industrial corporations. As an inevitable consequence, some crucial environmental pollution occurred in the 1950s and 1960s. In the rising concern over the problem, the government introduced many environmental protection laws in 1970 and established the Ministry of the Environment in 1971.

The oil crisis in 1973 also encouraged the efficient use of energy due to Japan's lack of natural resources. Current priority environmental issues include urban air pollution (NOx, suspended particulate matter, toxics), waste management, water eutrophication, nature conservation, climate change, chemical management and international co-operation for environmental conservation.

Today Japan is one of the world's leaders in the development of new environment-friendly technologies. Honda and Toyota hybrid electric vehicles were named to have the highest fuel economy and lowest emissions. This is due to the advanced technology in hybrid systems, biofuels, use of lighter weight material and better engineering.

As a signatory of the Kyoto Protocol, and host of the 1997 conference which created it, Japan is under treaty obligations to reduce its carbon dioxide emissions and to take other steps related to curbing climate change. The Cool Biz campaign introduced under former Prime Minister Junichiro Koizumi was targeted at reducing energy use through the reduction of air conditioning use in government offices. Japan is preparing to force industry to make big cuts in greenhouse gases, taking the lead in a country struggling to meet its Kyoto Protocol obligations. Japan is ranked 20th best in the world in the 2010 Environmental Performance Index.

## *Economy*

From 1868, the Meiji period launched economic expansion. Meiji rulers embraced the concept of a free market economy and adopted British and North American forms of free enterprise capitalism. Japanese went to study overseas and Western scholars were hired to teach in Japan. Many of today's enterprises were founded at the time. Japan emerged as the most developed nation in Asia.

From the 1960s to the 1980s, overall real economic growth has been called a "Japanese miracle": a 10% average in the 1960s, a 5% average in the 1970s and a 4% average in the 1980s. Growth slowed markedly in the 1990s during what the Japanese call the Lost Decade, largely because of the after-effects of Japanese asset price bubble and domestic policies intended to wring speculative excesses from the stock and real estate markets. Government efforts to revive economic growth met with little success and were further hampered by the global slowdown in 2000. The economy showed strong signs of recovery after 2005. GDP growth for that year was 2.8%, with an annualized fourth quarter expansion of 5.5%, surpassing the growth rates of the US and European Union during the same period.

As of 2009, Japan is the second largest economy in the world, after the United States, at around US$5 trillion in terms of nominal GDP and third after the United States and China in terms of purchasing power parity. Japan's public debt is 192 percent of its annual gross domestic product. Banking, insurance, real estate, retailing, transportation, telecommunications and construction are all major industries. Japan has a large industrial capacity and is home to some of the largest, leading and most technologically advanced producers of motor vehicles, electronic equipment, machine tools, steel and nonferrous metals, ships, chemicals, textiles and processed foods. The service sector accounts for three quarters of the gross domestic product.

As of 2001, Japan's shrinking labour force consisted of some 67 million workers. Japan has a low unemployment rate, around 4%. Almost one in six Japanese, or 20 million people, lived in poverty in 2007. Japan's GDP per hour worked is the world's 19th highest as of 2007. The Big Mac Index shows that Japanese workers get the highest salary per hour in the world. Some of the largest enterprises in Japan include Toyota, Nintendo, NTT DoCoMo, Canon, Honda, Takeda Pharmaceutical, Sony, Panasonic, Toshiba, Sharp, Nippon Steel, Tepco, Mitsubishi and 711. It is home to some of the world's largest banks, and the Tokyo Stock Exchange (known for its Nikkei 225 and Topix indices) stands as the second largest in the world by market capitalization. Japan is home to 326 companies from the Forbes Global 2000 or 16.3% (as of 2006).

Japan ranks 12th of 178 countries in the Ease of Doing Business Index 2008 and it has one of the smallest governments in the developed world. Japanese variant of capitalism has many distinct features. Keiretsu enterprises are influential. Lifetime employment and seniority-based career advancement are relatively common in Japanese work environment. Japanese companies are known for management methods such as "The Toyota Way". Shareholder activism is rare. Recently, Japan has moved away from some of these norms. In the Index of Economic Freedom, Japan is the 5th most laissez-faire of 30 Asian countries.

Japan's exports amounted to 4,210 U.S. dollars per capita in 2005. Japan's main export markets are the United States 22.8%, the European Union 14.5%, China 14.3%, South Korea 7.8%, Taiwan 6.8% and Hong Kong 5.6% (for 2006). Japan's main exports are transportation equipment, motor vehicles, electronics, electrical machinery and chemicals. Japan's main import markets are China 20.5%, U.S. 12.0%, the European Union 10.3%, Saudi Arabia 6.4%, UAE 5.5%, Australia 4.8%, South Korea 4.7% and Indonesia 4.2% (for 2006). Japan's main imports are machinery and equipment, fossil fuels, foodstuffs (in particular beef), chemicals, textiles and raw materials for its industries. By market share measures, domestic markets are the least open of any OECD country. Junichiro Koizumi administration commenced some pro-competition reforms and foreign investment in Japan has soared recently.

Japan's business culture has many indigenous concepts such as the nemawashi, the nenko system, the salaryman, and the office lady. Japan's housing market is characterized by limited land supply in urban areas. This is particularly true for Tokyo, the world's largest urban agglomeration GDP. More than half of Japanese live in suburbs or more rural areas, where detached houses are the dominant housing type. Agricultural businesses in Japan often utilize a system of terrace farming and crop yields are high. 13% of Japan's land is cultivated. Japan accounts for nearly 15% of the global fish catch, second only to China. Japan's agricultural sector is protected at high cost.

## *Science and Technology*

Japan is one of the leading nations in the fields of scientific research, particularly technology, machinery and biomedical research. Nearly 700,000 researchers share a US$130 billion research and development budget, the third largest in the world. Japan is a world leader in fundamental scientific research, having produced thirteen Nobel laureates in either physics, chemistry or medicine, three Fields medalists and one Gauss Prize laureate.

Some of Japan's more prominent technological contributions are found in the fields of electronics, automobiles, machinery, earthquake engineering, industrial robotics, optics, chemicals, semiconductors and metals. Japan leads the world in robotics production and use, possessing more than half (402,200 of 742,500) of the world's industrial robots used for manufacturing. It also produced QRIO, ASIMO and AIBO. Japan is the world's largest producer of automobiles and home to four of the world's fifteen largest automobile manufacturers and seven of the world's twenty largest semiconductor sales leaders as of today.

The Japan Aerospace Exploration Agency (JAXA) is Japan's space agency that conducts space and planetary research, aviation research, and development of rockets and satellites. It is a participant in the International Space Station and the Japanese Experiment Module (Kibo) was added to the International Space Station during Space Shuttle assembly flights in 2008. It has plans in space exploration, such as launching the *Venus Climate Orbiter* (*PLANET-C*) in 2010, developing the *Mercury Magnetospheric Orbiter* to be launched in 2013, and building a moonbase by 2030.

On September 14, 2007, it launched lunar orbit explorer "*SELENE*" (Selenological and Engineering Explorer) on an H-IIA (Model H2A2022) carrier rocket from Tanegashima Space Centre. *SELENE* is also known as *Kaguya*, the lunar princess of the ancient folktale *The Tale of the Bamboo Cutter*. *Kaguya* is the largest lunar probe mission since the Apollo program. Its mission is to gather data on the moon's origin and evolution. It entered into a lunar orbit on October 4, flying in a lunar orbit at an altitude of about 100 km (62 mi).

## Energy in Japan

Japan lacks significant domestic sources of fossil energy except coal and must import substantial amounts of crude oil, natural gas, and other energy resources, including uranium. In 1990 Japan's dependence on imports for primary energy stood at more than 85%, and the country had a total energy requirement of 428.2 million tons of petroleum equivalent.

### *Energy Use*

Japan's rapid industrial growth since the end of World War II doubled the nation's energy consumption every five years into the 1990s. During the 1960–72 period of accelerated growth, energy consumption grew much faster than GNP, doubling Japan's consumption of world energy. By 1976, with only 3% of the world's population, Japan was consuming 6% of global energy supplies.

In 1990, consumption totaled 298 million tons: 46.7% of which was used by industry; 23.3% by the transportation sector; 26.6% for agricultural, residential, services, and other uses; and 3.3% for non-energy uses, such as lubricating oil or asphalt.

### *Electricity Generation*

In 2008, Japan ranked third in the world in electricity production, after the United States and China, with 1.025 trillion kWh produced during that year.

In terms of per capita electricity consumption, the average person in Japan consumed 8,459 Kilowatt-Hours in 2004 compared to 14,240 for the average American. In that respect it ranked 18th among the countries of the world. Its per capita electricity consumption increased by 21.8% between 1990 and

2004. With 53 active nuclear power generating reactor units in 2009, Japan ranked third in the world in that respect, after the United States (104 reactors) and France (59). Almost one quarter (24.93) of its electricity production was from nuclear plants, compared to 76.18% for France and 19.66% for the United States.

In 1989 Japan was the world's third largest producer of electricity. About 75% of the available power was controlled by the ten major regional power utilities, of which Tokyo Electric Power Company was the world's largest. Electricity rates in Japan were among the world's highest.

### *Electrical Power Supply*

The standard voltage at power outlets is 100 V, but there are two frequencies in use: 50 Hz in Eastern Japan and 60 Hz in Western Japan.

### *Utilities*

In Japan, the electricity market is divided up into 10 regulated monopolies:

- Chugoku Electric Power Company (CEPCO)
- Chubu Electric Power (Chuden)
- Hokuriku Electric Power Company (HEPCO)
- Hokkaido Electric Power Company (Hokuden)
- Kyushu Electric Power (Kyuden)
- Kansai Electric Power Company (KEPCO)
- Okinawa Electric Power Company (Okiden)
- The Tokyo Electric Power Company (TEPCO)
- Tohoku Electric Power (Tohokuden)
- Shikoku Electric Power Company (Yonden).

### *Energy Supply*

In 1950 coal supplied half of Japan's energy needs, hydroelectricity one-third, and oil the rest. In 1988 oil provided Japan with 57.3% of energy needs, coal 18.1%, natural gas 10.1%, nuclear power 9.0%, hydroelectic power 4.6%, geothermal power 0.1%, and 1.3% came from other sources. By 2001 the contribution of oil had declined further to 50.2% of the total, with further rises in the use of nuclear power and natural gas.

### *Petroleum and Diversification*

After the two oil crises of the 1970s (1973 and 1979), the pattern of energy consumption in Japan changed from heavy dependence on oil to some diversification to other forms of energy resources in order to increase energy security. Japan's domestic oil consumption dropped slightly, from around 5.1 million barrels of oil per day in the late 1970s to 4.9 million barrels per day in 1990. While the country's use of oil declined, its consumption of nuclear power and LNG rose substantially. Several Japanese industries, including electric power companies and steelmakers, switched from petroleum to coal, most of which is imported.

The state stockpile equals about 92 days of consumption and the privately held stockpiles equal another 77 days of consumption for a total of 169 days or 579 million barrels. The Japanese SPR is run by the Japan Oil, Gas and Metals National Corporation.

### *Natural Gas*

Because domestic natural gas production is minimal, rising demand is met by greater imports. Japan's main LNG suppliers in 1987 were Indonesia (51.3%), Malaysia (20.4%), Brunei (17.8%), United Arab Emirates (7.3%), and the United States (3.2%).

### *Nuclear Power*

The Japanese were working to increase the availability of nuclear power in 1985. Although Japan was a late starter in this field, it finally imported technology from the United States and obtained uranium from Canada, France, South Africa, and Australia. By 1991 the country had 42 nuclear reactors in operation, with a total generating capacity of approximately 33 million kilowatts. The ratio of nuclear power generation to total electricity production increased from 2% in 1973 to 23.6% in 1990.

During the 1980s, Japan's nuclear power program was strongly opposed by environmental groups, particularly after the Three Mile Island accident in the United States. Other problems for the program were the rising costs of nuclear reactors and fuel, the huge investments necessary for fuel enrichment and reprocessing plants, reactor failures, and nuclear waste disposal. Nevertheless, Japan continued to build nuclear power plants.

### *Renewable Energy*

Of alternative energy sources, Japan has largely exploited geothermal energy. The country had six geothermal power stations with a combined capacity of 133,000 kilowatts in 1989.

In addition, although it only makes a minor contribution to the total, Japan was the world's second largest producer of photovoltaic electricity until overtaken by Germany in 2005, a year in which it had 38% of the world supply compared to Germany's 39%.

### *Carbon Emissions*

In 2003 Japan was the 5th largest producer of carbon emissions, generating 5% of the world total. In 2003 Japan ranked 36 in the list of countries by carbon dioxide emissions per capita.

Reports indicate Japan is having difficulty in meeting its 6% reduction target under the Kyoto Protocol in part because Japanese businesses are already very energy efficient. Despite this, former Prime Minister Shinzo Abe has called for a 50% cut in world emissions by 2050, and expects Japan to play a leading role in such an effort.

## Transport in Japan

Transportation in Japan is modern and infrastructure spending has been large.

Japan's road spending has also been large.. The 1.2 million kilometers of paved road are the main means of transportation. Japan has left-hand traffic. A single network of high-speed, divided, limited-access toll roads connects major cities and are operated by toll-collecting enterprises.

Dozens of Japanese railway companies compete in regional and local passenger transportation markets; for instance, 7 JR enterprises, Kintetsu Corporation, Seibu Railway, and Keio Corporation. Often, strategies of these enterprises contain real estate or department stores next to stations. Some 250 high-speed Shinkansen trains connect major cities. All trains are known for punctuality. There are 173 airports, and the largest domestic airport, Haneda Airport, is Asia's busiest airport. The largest international gateways are Narita International Airport (Tokyo area), Kansai International Airport, and Chubu Centrair International Airport (Nagoya area). The largest ports include Nagoya Port.

### *Rail Transportation*

In Japan, railways are a major means of passenger transportation, especially for mass and high-speed transport between major cities and for commuter transport in metropolitan areas. Seven Japan Railways Group companies, once state-owned until 1987, cover most parts of Japan. There also are railway services operated by private rail companies, regional governments, and companies funded by both regional governments and private companies. Japanese trains are also famous for always being on time. Five stations (Shinjuku Station, Ikebukuro Station, Shibuya Station, Umeda Station, and Yokohama Station) serve more than 2 million passengers each on an average day, making Japan the most railway using nation per capita.

Total railways of 23,670.7 km include entirely electrified 2,893.1 km of 1,435 mm standard gauge and 89.8 km of 1,372 mm (4 ft 6 in) narrow gauge, all of which is electrified. About the half of 20,656.8 km 1,067 mm (3 ft 6 in) gauge and 40 km of 762 mm (2 ft 6 in) gauge track are electrified.

Most Japanese people travelled on foot until the later part of the 19th century. The first railway was built between Tokyo and Yokohama in 1872 and many more developed. Japan now has one of the worlds most developed transportation networks. Mass transportation is well developed in Japan, but the road system lags behind and is inadequate for the number of cars. Road construction is difficult because of the high areas of population and the limited amount of usable land. Shinkansen are the high speed trains in Japan and they are known as bullet trains. About 250 Shinkansen trains operate daily. The fastest shinkansen trains are the N700 series and 500 series *Nozomi*, which operate at a maximum speed of 300 km/h. Shinkansen trains are known to be very punctual. A train is recorded as late if it does not arrive at the specified time. In

2003, the average delay per train on the Tokaido Shinkansen was 6 seconds.

### *Road Transportation*

Japan has 1,152,207 km of highways with 863,003 km (including 6,114 km of expressways) paved and 289,204 km of unpaved ways (1997 est.). A single network of high-speed, divided, limited-access toll roads connects major cities on Honshu, Shikoku and Kyushu. Hokkaido has a separate network, and Okinawa Island has a highway of this type. In the year 2005, the toll collecting companies, formerly Japan Highway Public Corporation, have been transformed into private companies in public ownership, and there are plans to sell parts of them. The aim of this policy is to encourage competition and decrease tolls.

Road passenger and freight transport expanded considerably during the 1980s as private ownership of motor vehicles greatly increased along with the quality and extent of the nation's roads. Bus companies including the JR Bus companies operates long-distance bus service on the nation's expanding expressway network. In addition to relatively low fares and deluxe seating, the buses are well utilized because they continue service during the night, when air and train service is limited.

The cargo sector grew rapidly in the 1980s, recording 274.2 billion tonne-kilometres in 1990. The freight handled by motor vehicles, mainly trucks, in 1990, was over 6 billion tonnes, accounting for 90 percent of domestic freight tonnage and about 50 percent of tonne-kilometres.

Recent large infrastructure projects were the construction of the Great Seto Bridge and the Tokyo Bay Aqua-Line (opened 1997).

Although road fatalities have been decreasing due in part to stricter enforcement of drink driving laws, 2004 still saw 7,358 deaths on Japanese roads.

### *Air Transport*

Japan has many airports. The main international gateways are Narita International Airport (Tokyo area), Kansai International Airport, and Chubu Centrair International Airport (Nagoya area). The main domestic hub is Tokyo International

Airport (Haneda Airport), Asia's busiest airport; other major traffic hubs include Osaka International Airport (Itami Airport), New Chitose Airport outside Sapporo, and Fukuoka Airport. 14 heliports are estimated to exist (1999).

The two main airlines are All Nippon Airways and Japan Airlines. Other passenger carriers include Skymark Airlines, Skynet Asia Airways, Air Do and Star Flyer. United Airlines and Delta Air Lines, formerly Northwest Airlines, are major international operators from Narita Airport.

Domestic air travel in Japan has historically been highly regulated. From 1972, the three major domestic airlines (JAL, ANA, and JAS) were allocated certain routes, with JAL and ANA sharing trunk routes, and ANA and JAS sharing local feeder routes. JAL also had a flag-carrier monopoly on international routes until 1986. Airfares were set by the government until 2000, although carriers had freedom to adjust the standard fares starting in 1995 (when discounts of up to 50% were permitted). Today, fares can be set by carriers, but the government retains the ability to veto fares that are impermissibly high.

Many airports are less busy than 'wildly over optimistic' initial forecasts.

### *Marine Transport*

There are 1770 km of Waterways in Japan; seagoing craft ply all coastal inland seas.

The twenty-two major seaports designated as special important ports by Ministry of Land, Infrastructure and Transport include Chiba, Fushiki/Toyama, Himeji, Hiroshima, Kawasaki, Kitakyushu, Kobe, Kudamatsu, Muroran, Nagoya, Niigata, Osaka, Sakai/Senpoku, Sendai/Shiogama, Shimizu, Shimonoseki, Tokyo, Tomakomai, Wakayama, Yokkaichi, and Yokohama.

Japan has 662 ships of with a volume of 1,000 gross register tons (GRT) or over, totaling 13,039,488 gross register tons (GRT) or 18,024,969 metric tons deadweight (DWT). There are 146 bulk ships, 49 cargo ships, 13 chemical tankers, 16 combination bulk, 4 with combination of ore and oil, 25 container, 45 liquefied gas, 9 passenger, 2 passenger and cargo combination ships, 214 petroleum tankers, 22 refrigerated cargo, 48 roll-on/roll-off ships, 9 short-sea passenger, and 60 vehicle carriers (1999 est.).

Ferries connect Hokkaido to Honshu, and Okinawa Island to Kyushu and Honshu. They also connect other smaller islands and the main islands. The scheduled international passenger routes are to China, Russia, South Korea and Taiwan. Coastal and cross-channel ferries on the main islands decreased in routes and frequencies following the development of bridges and expressways but some are still operating (as of 2007.)

### *Pipelines*

Japan has 84 km of pipelines for crude oil, 322 km for petroleum products, and 1,800 km for natural gas.

## Demographics

Japan's population is estimated at around 127.3 million. Japanese society is linguistically and culturally homogeneous with small populations of foreign workers. Zainichi Koreans, Zainichi Chinese, Filipinos, Japanese Brazilians, Japanese Peruvians are amongst the small minorities resident in Japan. In 2003, there were about 136,000 Western expatriates in Japan. The most dominant native ethnic group is the Yamato people; the primary minority groups include the indigenous Ainu and Ryukyuan, as well as social minority groups like the *burakumin*.

Japan has one of the highest life expectancy rates in the world, at 81.25 years of age as of 2006. The Japanese population is rapidly aging, the effect of a post-war baby boom followed by a decrease in births in the latter part of the twentieth century. In 2004, about 19.5% of the population was over the age of 65.

The changes in the demographic structure have created a number of social issues, particularly a potential decline in the workforce population and increases in the cost of social security benefits such as the public pension plan. Many Japanese youth are increasingly preferring not to marry or have families as adults. Japan's population is expected to drop to 100 million by 2050 and to 64 million by 2100. Demographers and government planners are currently in a heated debate over how to cope with this problem. Immigration and birth incentives are sometimes suggested as a solution to provide younger workers to support the nation's aging population. According to the UNHCR, in 2007 Japan accepted just 41 refugees for resettlement, while the United States took in 50,000.

Japan suffers from a high suicide rate. In 2009, the number of suicides exceeded 30,000 for the twelfth straight year. Suicide is the leading cause of death for people under 30.

### *Religion in Japan*

The highest estimates for the number of Buddhists and Shintoists in Japan is 84–96%, representing a large number of believers in a syncretism of both religions. However, these estimates are based on people with an association with a temple, rather than the number of people truly following the religion. Professor Robert Kisala (Nanzan University) suggests that only 30 percent of the population identify themselves as belonging to a religion.

Taoism, Confucianism and Buddhism from China have also influenced Japanese beliefs and customs. Religion in Japan tends to be syncretic in nature, and this results in a variety of practices, such as parents and children celebrating Shinto rituals, students praying before exams, couples holding a wedding at a Christian church and funerals being held at Buddhist temples. A minority (2,595,397, or 2.04%) profess to Christianity. In addition, since the mid-19th century, numerous religious sects (*Shinshukyo*) have emerged in Japan, such as Tenrikyo and Aum Shinrikyo (or Aleph).

### *Languages in Japan*

More than 99% of the population speaks Japanese as their first language. It is an agglutinative language distinguished by a system of honorifics reflecting the hierarchical nature of Japanese society, with verb forms and particular vocabulary which indicate the relative status of speaker and listener. According to a Japanese dictionary *Shinsen-kokugojiten*, Chinese-based words make up 49.1% of the total vocabulary, indigenous words are 33.8% and other loanwords are 8.8%.

The writing system uses kanji (Chinese characters) and two sets of kana (syllabaries based on simplified Chinese characters), as well as the Latin alphabet and Arabic numerals.

The Ryukyuan languages, also part of the Japonic language family to which Japanese belongs, are spoken in Okinawa, but few children learn these languages. The Ainu language is moribund, with only a few elderly native speakers remaining in Hokkaido. Most public and private schools require students to take courses in both Japanese and English.

## *Education and Health*

Primary, secondary schools and universities were introduced into Japan in 1872 as a result of the Meiji Restoration. Since 1947, compulsory education in Japan consists of elementary school and middle school, which lasts for nine years (from age 6 to age 15). Almost all children continue their education at a three-year senior high school, and, according to the MEXT, about 75.9% of high school graduates attend a university, junior college, trade school, or other post-secondary institution in 2005.

Japan's education is very competitive, especially for entrance to institutions of higher education. The two top-ranking universities in Japan are the University of Tokyo and Kyoto University. The Programme for International Student Assessment coordinated by the OECD, currently ranks Japanese knowledge and skills of 15-year-olds as the 6th best in the world.

In Japan, healthcare services are provided by national and local governments. Payment for personal medical services is offered through a universal health care insurance system that provides relative equality of access, with fees set by a government committee. People without insurance through employers can participate in a national health insurance program administered by local governments. Since 1973, all elderly persons have been covered by government-sponsored insurance. Patients are free to select physicians or facilities of their choice.

## Culture and Recreation

Japanese culture has evolved greatly over the years, from the country's original Jomon culture to its contemporary culture, which combines influences from Asia, Europe and North America. Traditional Japanese arts include crafts (ikebana, origami, ukiyo-e, dolls, lacquerware, pottery), performances (bunraku, dance, kabuki, noh, rakugo), traditions (games, tea ceremony, Budo, architecture, gardens, swords) and cuisine. The fusion of traditional woodblock printing and Western art led to the creation of manga, a typically Japanese comic book format that is now popular within and outside Japan. Manga-influenced animation for television and film is called anime. Japanese-made video game consoles have prospered since the 1980s.

Japanese music is eclectic, having borrowed instruments, scales and styles from neighbouring cultures. Many instruments, such as the koto, were introduced in the ninth and tenth centuries. The accompanied recitative of the Noh drama dates from the fourteenth century and the popular folk music, with the guitar-like shamisen, from the sixteenth. Western classical music, introduced in the late nineteenth century, now forms an integral part of the culture. Post-war Japan has been heavily influenced by American and European modern music, which has led to the evolution of popular band music called J-pop.

Karaoke is the most widely practiced cultural activity. A November 1993 survey by the Cultural Affairs Agency found that more Japanese had sung karaoke that year than had participated in traditional cultural pursuits such as flower arranging or tea ceremonies.

The earliest works of Japanese literature include two history books the *Kojiki* and the *Nihon Shoki* and the eighth century poetry book *Manyoshu*, all written in Chinese characters. In the early days of the Heian period, the system of transcription known as *kana* (Hiragana and Katakana) was created as phonograms. *The Tale of the Bamboo Cutter* is considered the oldest Japanese narrative. An account of Heian court life is given by *The Pillow Book* written by Sei Shonagon, while *The Tale of Genji* by Lady Murasaki is often described as the world's first novel.

During the Edo period, literature became not so much the field of the samurai aristocracy as that of the chonin, the ordinary people. Yomihon, for example, became popular and reveals this profound change in the readership and authorship. The Meiji era saw the decline of traditional literary forms, during which Japanese literature integrated Western influences.

Natsume Soseki and Mori Ogai were the first "modern" novelists of Japan, followed by Ryunosuke Akutagawa, Junichiro Tanizaki, Yasunari Kawabata, Yukio Mishima and, more recently, Haruki Murakami. Japan has two Nobel Prize-winning authors—Yasunari Kawabata (1968) and Kenzaburo Oe (1994).

### *Sports*

Traditionally, sumo is considered Japan's national sport and it is a popular spectator sport. Martial arts such as judo, karate and modern kendo are also widely practiced and enjoyed by spectators in the country. After the Meiji Restoration, many Western sports were introduced in Japan and began to spread through the education system.

The professional baseball league in Japan was established in 1936. Today baseball is the most popular spectator sport in the country. One of the most famous Japanese baseball players is Ichiro Suzuki, who, having won Japan's Most Valuable Player award in 1994, 1995, and 1996, now plays for the Seattle Mariners of North American Major League Baseball. Prior to that, Sadaharu Oh was well-known outside Japan, having hit more home runs during his career in Japan than his contemporary, Hank Aaron, did in the United States.

Since the establishment of the Japan Professional Football League in 1992, association football (soccer) has also gained a wide following. Japan was a venue of the Intercontinental Cup from 1981 to 2004 and co-hosted the 2002 FIFA World Cup with South Korea. Japan is one of the most successful soccer teams in Asia, winning the Asian Cup three times.

Golf is also popular in Japan, as are forms of auto racing, such as the Super GT sports car series and Formula Nippon formula racing. Twin Ring Motegi was completed in 1997 by Honda in order to bring IndyCar racing to Japan.

Japan hosted the Summer Olympics in Tokyo in 1964. Japan has hosted the Winter Olympics twice: Nagano in 1998 and Sapporo in 1972.

### *Ryukyu Islands*

The Ryukyu Islands, also known as the Nansei Islands, is a chain of islands in the western Pacific, on the eastern limit of the East China Sea and to the southwest of the island of

Kyushu in Japan. From about 1829 until the mid 20th century, they were alternately called Luchu, Loochoo, or Lewchew, akin to the Mandarin pronunciation *Liuqiu*. They stretch southwest from the Japanese island of Kyushu to within 120 kilometres (75 mi) of the island of Taiwan.

The islands are administratively divided into the Satsunan Islands to the north, belonging to Kagoshima Prefecture, and Ryukyu Shoto to the south, belonging to Okinawa Prefecture, Japan. Yoron Island is the southernmost island of the Satsunan Islands, and Yonaguni is the southernmost of the Ryukyu Islands. The largest of the islands is Okinawa Island.

The islands have a subtropical climate with warm winters and hot summers. Precipitation is very high, and is affected by the rainy season and typhoons.

The archipelago is home to the Ryukyuan languages. The original dialects are native to each island and distinct from one another.

On February 27, 2010, at 5:31 a.m. local time, a magnitude 7.0 earthquake occurred 80 kilometres (50 mi) east-southeast of Naha at a depth of 22 kilometres (14 mi), but no major damage was reported. At least eight recorded aftershocks were reported, with magnitude up to 5.3

## Naming

### *Japanese*

In Japanese, the definition of the *Ryukyu Shoto*, literally meaning "Ryukyu Islands", is somewhat different from the English definition of the word. In Japanese, the term *Ryukyu Shoto* is used to refer to the part of the Nansei Islands which is in Okinawa Prefecture (the southern half), as opposed to islands of the same group located in Kagoshima Prefecture (the northern half).

Modern usage of the word *Ryukyu* in Japanese, however, is usually replaced by the word *Okinawa*, which is considered its synonym. When referring to the region in the broad sense, the Nansei Islands are sometimes referred to as *Amami-Okinawa Chiho*, literally "Amami-Okinawa Region", or variations thereof. For example, the Japanese train timetables

*JR Jikokuhyo* uses variations of *Nansei Shoto*, *Okinawa*, *Amami*, etc., but completely avoids using the word *Ryukyu*.

### *English*

In English, until well into the late 1800s (Meiji period in Japan), the word "Ryukyu" was spelled *Luchu, Loo-choo,* or *Lewchew*. These spellings were based on the Chinese pronunciation of the characters for "Ryukyu", which in Mandarin is *Liuqiu*.

### *History*

The Ryukyu Kingdom was once an independent kingdom occupying the island chain, from Yonaguni Island in the southwest to Amami Oshima in the north. In 1372, it became a tributary state of the Ming Dynasty.

In 1609, Shimazu Tadatsune, Lord of Satsuma, invaded the Ryukyu Kingdom with a fleet of 13 junks and 2,500 samurai, thereby establishing suzerainty over the islands. They faced little opposition from the Ryukyuans, who lacked any significant military capabilities, and who were ordered by King Sho Nei to surrender peacefully rather than suffer the loss of precious lives. After that, the kings of the Ryukyus paid tribute to the Japanese shogun as well as the Chinese emperor.

In 1879, the Meiji government announced the annexation of the Ryukyus. Messengers sent by the Ryukyuan king had knelt outside the Zongli Yamen, the Chinese foreign affairs office in Beijing, for three days, pleading not to be separated from China. China, weakened from internal corruption and colonial occupation, refused the request to send military protection. Instead, China made diplomatic objections and asked former United States President Ulysses S. Grant to arbitrate. Grant decided that Japan's claim to the islands was stronger and ruled in Japan's favour. The claims of the indigenous Ryukyuans to the land were ignored.

In the process of annexation, the Japanese military assassinated Ryukyu politicians and civilians who opposed the takeover. The Ryukyu Kingdom became part of its northern neighbor, the Satsuma han. Later, it became its own prefecture, Okinawa Prefecture, when the prefectural system was adopted nationwide. Compulsory Japanese education was enforced on the Ryukyu children, whereby they were taught Japanese language, culture and identity, while strictly forbidden the use of their native language.

Military activity on the island, before and during World War II, especially the Battle of Okinawa, had a devastating effect on the Okinawan people. A huge loss of civilian life left many feeling that they were being mistreated by both the Japanese and American military. Okinawa remains the poorest prefecture in Japan to this day.

The US was granted control over Ryukyu Islands south of 29°N latitude amongst other Pacific islands, under the San Francisco Peace Treaty between the Allied Powers and Japan. US military control over Okinawa began in 1945 with establishment of the Okinawa Advisory Council. This organization eventually became the government of the Ryukyu Islands which existed from 1952 to 1972. Sovereignty was given to Japan in 1972.

Today, there are a number of issues arising from Ryukyuan history. Some Ryukyuans and some Japanese feel that people from the Ryukyus are different from the majority Yamato people. Some natives of the Ryukyus claim that the central government is discriminating against the islanders by allowing so many American soldiers to be stationed on bases in Okinawa with a minimal presence on the mainland. Additionally, there is some discussion of secession from Japan.

Many popular singers and musical groups come from the Ryukyus. These include (among many others) the pop groups Begin and Orange Range, singers Namie Amuro and Gackt, as well as the group Da Pump. See also Ryukyuan music.

### *Historical Description of the* 'Loo-Choo' *Islands*

An article in the 1878 edition of the *'Globe Encyclopaedia of Universal Information'* described the islands as:

Loo-Choo, Lu-Tchu, or Lieu-Baeu, a group of thirty-six islands stretching from Japan to Formosa, in 20°-27° 40' N. lat., 126" :o'-!29° 5' E. long., and tributary to Japan. The largest, Tsju San ('middle island') is about 60 miles long and 12 broad; others are Sannan in the S. and Sanbok in the N. Nawa, the chief port of Tsju San, is open to foreign commerce. The islands enjoy a magnificent climate, and are highly cultivated and very productive. Among the productions are tea, rice, sugar, tobacco, camphor, fruits, and silk. The principal manufactures are cotton, paper, porcelain, and lacquered ware. The people, who are small, seem a connecting link between the Chinese and Japanese.

### *People*

The Ryukyuans are known for their longevity. The Okinawa Centenarian Study attributes this phenomenon to a combination of diet, exercise, and lifestyle practices.

Since the most recent Japanese invasion in 1879, Japanese has become the main language on the Ryukyus, especially on Uchina (Okinawa), through discrimination in education. Middle-aged or younger people tend not to speak a Ryukyuan language as fluently as Japanese, if at all.

## Ecology

### *Nansei Islands Subtropical Evergreen Forests*

The Ryukyu Islands are recognized by ecologists as a distinct subtropical moist broadleaf forest ecoregion. The flora and fauna of the islands have much in common with Taiwan, the Philippines, and Southeast Asia, and are part of the Indomalaya ecozone.

### *Coral Reefs*

The coral reefs of the Ryukyus are one of the World Wildlife Fund's Global 200 ecoregions. The reefs are endangered by sedimentation and eutrophication, mostly a result of agriculture, as well as damage from fishing.

### *Major Islands*

This list is based on present day Japanese geographic names:

- Nansei Islands
    - Satsunan Islands (The Northern Half)
        - Osumi Islands with:
        - Tanegashima, Yakushima, Kuchinoerabujima, Mageshima in the North-Eastern Group,

  * Takeshima, Iojima, Kuroshima in the North-Western Group.
  * Tokara Islands (The *Shichi-to*): Kuchinoshima, Nakanoshima, Gajajima, Suwanosejima, Akusekijima, Tairajima, Kodakarajima, Takarajima
  * Amami Islands: Amami Oshima, Kikaigashima, Kakeromajima, Yoroshima, Ukeshima, Tokunoshima, Okinoerabujima, Yoronjima

- o Ryukyu Shoto (The Southern Half)
  * Okinawa Islands (The Central Group or Ryukyu proper): Okinawa Island (aka. Okinawan mainland, Okinawa honto), Kumejima, Iheyajima, Izenajima, Agunijima, Iejima, Iwo Tori Shima (Iotorishima)
  * Kerama Islands: Tokashikijima, Zamamijima, Akajima, Gerumajima
  * Daito Islands: Kita Daito, Minami Daito, Oki Daito
  * Sakishima Islands ("Further Isles")
  * Miyako Islands: Miyakojima, Ikema, Ogami, Irabu, Shimoji, Kurima, Minna, Tarama
  * Yaeyama Islands: Iriomote, Ishigaki, Taketomi, Kohama, Kuroshima, Aragusuku, Hatoma, Yubujima, Hateruma, Yonaguni
  * Senkaku Islands (claimed by PRC and ROC): Uotsurijima, Kuba Jima, Taisho Jima, Kita Kojima, Minami Kojima.

# 6

# The South-West Pacific—Rich Tropical Islands

The Pacific Ocean is the largest of the Earth's oceanic divisions. It extends from the Arctic in the north to the Southern Ocean in the south, bounded by Asia and Australia in the west, and the Americas in the east.

At 169.2 million square kilometres (65.3 million square miles) in area, this largest division of the World Ocean – and, in turn, the hydrosphere – covers about 46% of the Earth's water surface and about 30% of its total surface. The equator subdivides it into the North Pacific Ocean and South Pacific Ocean, with two exceptions: the Galapagos and Gilbert Islands, while straddling the equator, are deemed wholly within the South Pacific. The Mariana Trench in the western North Pacific is the deepest point in the Pacific and in the world, reaching a depth of 10,911 metres (35,797 ft).

The Pacific Ocean was sighted by Europeans early in the 16th century, first by the Spanish explorer Vasco Nunez de Balboa who crossed the Isthmus of Panama in 1513 and named it *Mar del Sur* (South Sea). Its current name is however derived from the Luso-Latin macaronic *Tepre Pacificum*, "peaceful sea", bestowed upon it by the Portuguese explorer Ferdinand Magellan

## Overview

The Pacific Ocean encompasses approximately one-third of the Earth's surface, having an area of 179.7 million square kilometres (69.4 million sq mi and 161 million cubic mi) — significantly larger than Earth's entire landmass, with room for another Africa to spare. Extending approximately 15,500 kilometres (9,600 mi) from the Bering Sea in the Arctic to the northern extent of the circumpolar Southern Ocean at 60° S (older definitions extend it to Antarctica's Ross Sea), the Pacific reaches its greatest east-west width at about 5°N latitude, where it stretches approximately 19,800 kilometres (12,300 mi) from Indonesia to the coast of Colombia and Peru – halfway across the world, and more than five times the diameter of the Moon. The lowest known point on earth—the Mariana Trench—lies 10,911 metres (35,797 ft) below sea level. Its average depth is 4028~4188metres (14,000 ft).

The Pacific contains about 25,000 islands (more than the total number in the rest of the world's oceans combined), the majority of which are found south of the equator. Including partially submerged islands, the figure is substantially higher.

The Pacific Ocean is currently shrinking from plate tectonics, while the Atlantic Ocean is increasing in size, by roughly an inch per year (2–3 cm/yr) on 3 sides, roughly averaging 0.2 square miles (0.5 km) a year.

Along the Pacific Ocean's irregular western margins lie many seas, the largest of which are the Celebes Sea, Coral Sea, East China Sea, Philippine Sea, Sea of Japan, South China Sea, Sulu Sea, Tasman Sea, and Yellow Séa. The Strait of Malacca joins the Pacific and the Indian Oceans on the west, and Drake Passage and the Straits of Magellan link the Pacific with the Atlantic Ocean on the east. To the north, the Bering Strait connects the Pacific with the Arctic Ocean.

As the Pacific straddles the 180th meridian, the *West Pacific* (or *western Pacific*, near Asia) is in the Eastern Hemisphere, while the *East Pacific* (or *eastern Pacific*, near the Americas) is in the Western Hemisphere. For most of Magellan's voyage from the Strait of Magellan to the Philippines, the explorer indeed found the ocean peaceful. However, the Pacific is not always peaceful. Many tropical storms batter the islands of the Pacific. The lands around the Pacific Rim are full of volcanoes and often affected by earthquakes. Tsunamis, caused by underwater earthquakes, have devastated many islands and destroyed entire towns.

### *Extent*

The 3rd edition of the International Hydrographic Organization's (IHO) *Limits of Oceans and Seas* defines the limits of the North Pacific Ocean as follows:

- *On the Southwest.* The Northeastern limit of the East Indian Archipelago from the Equator to Luzon Island.
- *On the West and Northwest.* The Eastern limits of the Philippine Sea and Japan Sea and the Southeastern limit of the Sea of Okhotsk.
- *On the North.* The Southern limits of the Bering Sea and the Gulf of Alaska.
- *On the East.* The Western limit of Coastal waters of Southeast Alaska and Br. Columbia, and the Southern limit of the Gulf of California.
- *On the South.* The Equator, but excluding those islands of the Gilbert and Galapagos Groups which lie to the Northward thereof.

And the South Pacific Ocean is defined as:

- *On the West.* From Southeast Cape, the Southern point of Tasmania, down the meridian of 146°55'E to the Antarctic continent.
- *On the Southwest and Northwest.* The Southern, Eastern and Northeastern limits of the Tasman Sea, the Southeastern and Northeastern limits of the Coral Sea, the Southern, Eastern and Northern limits of the Solomon and Bismark seas, and the Northeastern limit of the East Indian Archipelago from New Guinea to the Equator.
- *On the North.* The Equator, but including those islands of the Gilbert and Galapagos Groups which lie to the Northward thereof.
- *On the East.* The meridian of Cape Horn (67°16'W) from Tierra del Fuego to the Antarctic continent; a line from Cape Virgins to Cape Espititu Santo, Tierra del Fuego, the Eastern entrance to Magellan Strait. (These limits have not yet been officially accepted by Argentina and Chile.)
- *On the South.* The Antarctic continent.

Note that these definitions exclude any marginal waterbodies that are separately defined by the International Hydrographic Organization (such as the Gulf of Alaska and Coral Sea), though these are usually considered to be part of the Pacific Ocean.

In 2000 the IHO redefined the Pacific Ocean, moving its southern limit to 60°S, with the waters south of that line identified as the Southern Ocean. This new definition has not yet been ratified (a reservation has been lodged by Australia) though it is in use by the IHO and others. If and when adopted, the 2000 definition will be published in the 4th edition of *Limits of Oceans and Seas*, restoring the Southern Ocean as originally outlined in the 2nd edition and subsequently omitted from the 3rd edition.

### *Water Characteristics*

The volume of the Pacific Ocean is approximately 622 million cubic km. Water temperatures in the Pacific vary from freezing in the poleward areas to about 30 °C (86 °F) near the equator. Salinity also varies latitudinally. The water near the equator is less salty than that found in the mid-latitudes because of abundant equatorial precipitation throughout the year. Poleward of the temperate latitudes salinity is also low, because little evaporation of seawater takes place in these frigid areas.

The motion of Pacific waters is generally clockwise in the Northern Hemisphere (the North Pacific gyre) and counter-clockwise in the Southern Hemisphere. The North Equatorial Current, driven westward along latitude 15°N by the trade winds, turns north near the Philippines to become the warm Japan or Kuroshio Current.

Turning eastward at about 45°N, the Kuroshio forks and some waters move northward as the Aleutian Current, while the rest turn southward to rejoin the North Equatorial Current. The Aleutian Current branches as it approaches North America and forms the base of a counter-clockwise circulation in the Bering Sea. Its southern arm becomes the chilled slow, south-flowing California Current.

The South Equatorial Current, flowing west along the equator, swings southward east of New Guinea, turns east at about 50°S, and joins the main westerly circulation of the

Southern Pacific, which includes the Earth-circling Antarctic Circumpolar Current. As it approaches the Chilean coast, the South Equatorial Current divides; one branch flows around Cape Horn and the other turns north to form the Peru or Humboldt Current.

## *Geology*

The andesite line is the most significant regional distinction in the Pacific. It separates the deeper, mafic igneous rock of the Central Pacific Basin from the partially submerged continental areas of felsic igneous rock on its margins. The andesite line follows the western edge of the islands off California and passes south of the Aleutian arc, along the eastern edge of the Kamchatka Peninsula, the Kuril Islands, Japan, the Mariana Islands, the Solomon Islands, and New Zealand's North Island.

The dissimilarity continues northeastward along the western edge of the Andes Cordillera along South America to Mexico, returning then to the islands off California. Indonesia, the Philippines, Japan, New Guinea, and New Zealand lie outside the Andesite Line.

The Pacific Ocean takes up roughly one third of the Earth's surface, having an area of 179.7 million square kilometres. It contains about 25,000 islands (more than the total number in the rest of the world's oceans combined), the majority of which are found south of the equator, Including partially submerged islands. The ocean was mapped by a man named Abraham Ortelius, he called it Maris Pacifici because of Ferdinand Magellan, who sailed the Pacific during his circumnavigation from 1519 to 1522 and said that it was much more calm than the Atlantic.The Pacific Ocean has several long seamount chains (which are chains of mountains submerging from the ocean seafloor) formed by hotspot volcanism. "The Ring of Fire" is the world's largest belt of explosive volcanism. It is named after the several hundred active volcanoes that sit above the various subduction zones (which is a geological process in which one edge of a crustal plate is forced sideways and downward into the mantle below another plate).

Within the closed loop of the Andesite Line are most of the deep troughs, submerged volcanic mountains, and oceanic volcanic islands that characterize the Pacific basin. Here basaltic lavas gently flow out of rifts to build huge dome-shaped volcanic mountains whose eroded summits form island arcs, chains, and clusters. Outside the Andesite Line, volcanism is of the explosive type, and the Pacific Ring of Fire is the world's foremost belt of explosive volcanism. The Ring of Fire is named after the several hundred active volcanoes that sit above the various subduction zones.

The Pacific Ocean is the only ocean which is almost totally bounded by subduction zones. Only the Antarctic and Australian coasts have no nearby subduction zones.

### *Seamount Chains*

The Pacific Ocean contains several long seamount chains, formed by hotspot volcanism. These include the Hawaiian – Emperor seamount chain chain and the Louisville seamount chain.

## Landmasses

The largest landmass entirely within the Pacific Ocean is the island of New Guinea— the second largest island in the world. Almost all of the smaller islands of the Pacific lie between 30°N and 30°S, extending from Southeast Asia to Easter Island; the rest of the Pacific Basin is almost entirely submerged. During the Last glacial period, New Guinea was part of Australia so the largest landmass would have been Borneo–Palawan.

The great triangle of Polynesia, connecting Hawaii, Easter Island, and New Zealand, encompasses the island arcs and clusters of the Cook Islands, Marquesas Islands, Samoa, Society, Tokelau, Tonga, Tuamotu, Tuvalu and the Wallis and Futuna islands.

North of the equator and west of the International Date Line are the numerous small islands of Micronesia, including the Caroline Islands, the Marshall Islands and the Mariana Islands.

In the southwestern corner of the Pacific lie the islands of Melanesia, dominated by New Guinea. Other important island groups of Melanesia include the Bismarck Archipelago, Fiji, New Caledonia, the Solomon Islands and Vanuatu. Islands in the Pacific Ocean are of four basic types: continental islands, high islands, coral reefs, and uplifted coral platforms. Continental islands lie outside the Andesite line and include New Guinea, the islands of New Zealand, and the Philippines. These islands are structurally associated with nearby continents. High islands are of volcanic origin, and many contain active volcanoes. Among these are Bougainville, Hawaii, and the Solomon Islands.

The third and fourth types of islands are both the result of coralline island building. Coral reefs are low-lying structures that have built up on basaltic lava flows under the ocean's surface. One of the most dramatic is the Great Barrier Reef off northeastern Australia. A second island type formed of coral is the uplifted coral platform, which is usually slightly larger than the low coral islands. Examples include Banaba (formerly Ocean Island) and Makatea in the Tuamotu group of French Polynesia.

## Pacific Islands

The Pacific Islands comprise 20,000 to 30,000 islands in the Pacific Ocean. Those islands lying south of the tropic of Cancer are traditionally grouped into three divisions: Melanesia, Micronesia and Polynesia.

### *Oceania*

Pacific islands are also sometimes collectively called *Oceania* (although Oceania is sometimes defined as also including Australasia and the Malay Archipelago).

- Melanesia means *black islands*. These include New Guinea (the largest Pacific island, which is divided into the sovereign nation of Papua New Guinea and the Indonesian provinces of Maluku, Papua and West Papua), New Caledonia, Zenadh Kes (Torres Strait Islands), Vanuatu, Fiji, and the Solomon Islands.
- Micronesia means *small islands*. These include the Marianas, Guam, Wake Island, Palau, the Marshall Islands, Kiribati, *Nauru, and the Federated States of Micronesia. Most of these lie north of the equator.
- Polynesia means *many islands*. These include New Zealand, the Hawaiian Islands, Rotuma, the Midway Islands, Samoa, American Samoa, Tonga, Tuvalu, the

Cook Islands, Wallis and Futuna, Tokelau, Niue, French Polynesia, and Easter Island. It is the largest of the three zones.

The region's islands are classified into two groups, high islands and low islands. Volcanoes form high islands, which generally can support more people and have a more fertile soil. Low islands are reefs or atolls, and are relatively small and infertile. Melanesia, the most populous of the three regions, contains mainly high islands, while most of Micronesia and Polynesia are low islands.

In addition, there are many other islands located within the boundaries of the Pacific Ocean that are not considered part of Oceania. These islands include the Galapagos Islands of Ecuador; the Aleutian Islands in Alaska; the Russian islands of Sakhalin and Kuril Islands; Taiwan; the Philippines; the South China Sea Islands; most of the islands of Indonesia; and the island nation of Japan, which includes the Ryukyu Islands.

Nauru (along with Kiribati's Banaba island) could be counted as somewhat of an exception. The indigenous Nauruans are both a mosaic and mixture of groups from all three categories-with cultural influence stemming primarily from Micronesia. The island was also said to be an extreme point of the "Tongan Empire" and may as a result share subtle cultural and, obviously, historical aspects with Polynesia. Lastly, the people speak a language and have a number of genes not in common with any of the three regions. Of the three, Nauru is least like Polynesia and Melanesia and for practical applications, Nauru is either assigned to Micronesia or designated as a separate entity (with the former being the most common).

## Oceania

Oceania (sometimes Oceanica) is a geographical, and often geopolitical, region consisting of numerous lands—mostly islands in the Pacific Ocean and vicinity. The term "Oceania" was coined in 1831 by French explorer Dumont d'Urville. The term is also sometimes used to denote a continent comprising Australia Continent and proximate Pacific islands, and is one of eight terrestrial ecozones. The boundaries of Oceania are defined in a number of ways. Most definitions include Australia, New Zealand and all or part of the Malay Archipelago.

Ethnologically, the islands that are included in Oceania are divided into the subregions of Melanesia, Micronesia, and Polynesia.

## *Extent*

Oceania is traditionally understood as being composed of three regions: Micronesia, Melanesia and Polynesia. As with any region, however, interpretations vary; increasingly, geographers and scientists divide Oceania into Near Oceania and Remote Oceania.

Most of Oceania consists of island nations comprising thousands of coral atolls and volcanic islands, with small human populations. Australia is the only continental country but Indonesia has land borders with Papua New Guinea, East Timor, and Malaysia. If the Australia-New Guinea continent is included then the highest point is Puncak Jaya in Papua at 4,884 m (16,024 ft) and the lowest point is Lake Eyre, Australia at 16 m (52 ft) below sea level.

## *Territories and Regions*

Descriptions of the regions and constituents of Oceania vary according to source. The information shown follows sources in cross-referenced articles; where sources differ, provisos have been clearly indicated. These territories and regions are subject to various additional categorisations, of course, depending on the source and purpose of each description.

## *Interpretative Details and Controversies*

- New Zealand is the western corner of the Polynesian Triangle. Its indigenous Maori constitute one of the major cultures of Polynesia. It is also, however, considered part of Australasia. More restricted definitions of the region may exclude New Zealand.
- Hawaii is the northern corner of the Polynesian Triangle and is generally included in Oceania, though politically it is part of the United States. The Hawaiian language is a Polynesian member of the Oceanic language family, and Hawaiian culture is one of the major cultures of Polynesia.
- The US territories in the North Pacific are generally considered part of Oceania.

- Rapa Nui, or Easter Island, is the eastern corner of the Polynesian triangle. A Polynesian island in the eastern Pacific Ocean and part of the territory of Chile, it is generally included in Oceania, in which case the most easterly place in Polynesia and Oceania is its dependency Isla Salas y Gómez 415 km to the East.
- The line in Indonesia dividing Oceania from Asia varies in location and is sometimes considered to be the Wallace Line. See the transcontinental country article.
- East Timor is often reckoned as a part of Oceania due to its location to the east of the Wallace Line and its cultural ties to Pacific peoples. Biogeographically, East Timor lies within Wallacea, an ecological transition zone between Asia and Australasia. This transition is less known and less favoured these days as a continental boundary.
- Australia is sometimes *not* included in Oceania. Terms such as *Pacific Islands* or *South Sea Islands* might be used to describe Oceania without Australia (and New Zealand). The term "Australasia" invariably includes Australia, and usually includes New Zealand, Papua New Guinea, and some other parts of Oceania. This term is sometimes controversial, though, as it may be interpreted as implying an association with Asia — a separate continent — or too great an association with Australia. The term is actually derived from the word *"Austral"*, meaning "of, relating to, or coming from the south". This word represents the common root of both names: Australia and Australasia.
- Although Christmas Island and the Cocos (Keeling) Islands belong to the Commonwealth of Australia, they are nearer Indonesia than the Australian mainland, and are commonly associated with Asia instead of Oceania.
- In its widest sense, the term may embrace the entire insular region between Asia and the Americas, thereby including other islands in the Pacific Rim such as the Ryukyu, Kuril and Aleutian islands, the Japanese Archipelago and Taiwan.

- Amateur radio defines the continental boundaries somewhat differently. The Worked All Continents award includes all of Indonesia and the Philippines in Oceania, places Easter Island with Chile, and makes some other minor changes.

### *Ecogeography*

Oceania is one of eight terrestrial ecozones, which constitute the major ecological regions of the planet. The Oceania ecozone includes all of Micronesia, Fiji, and all of Polynesia except New Zealand. New Zealand, New Guinea and nearby islands, Australia, the Solomon Islands, Vanuatu, and New Caledonia constitute the separate Australasia ecozone.

### *Religion*

Religion in Oceania remains dominated demographically by Christianity. Traditional religions are often animist and prevalent among traditional tribes is the belief in evil spirits (*masalai* in Tok Pisin), which are blamed for "poisoning" people, causing calamity and death. In recent Australian and New Zealand censuses, large proportions of the population say they belong to "No religion" (which includes humanism, atheism, agnosticism, and rationalism). In Tonga, everyday life is heavily influenced by Polynesian traditions and especially by the Christian faith. The Bahai House of Worship in Tiapapata, Samoa is one of seven designations administered in the Bahai faith.

## Sport

### *Pacific Games*

The Pacific Games (formerly known as the South Pacific Games) is a multi-sport event, much like the Olympics, (albeit on a much smaller scale), with participation exclusively from countries around the Pacific. It is held every four years and began in 1963.

### *Rugby League*

Rugby league is a popular sport throughout Oceania. Rugby League is the national sport in Papua New Guinea (the second most populous country in Oceania after Australia) and is very popular in Australia and attracts significant attention across

New Zealand and the Pacific Islands. Australia and New Zealand are two of the best sides in the world. Australia has won the Rugby League World Cup a record nine times while New Zealand won their first World Cup in 2008. Australia hosted the second tournament in 1957. Australia and New Zealand jointly hosted it in 1968 and 1977. New Zealand hosted the final for the first time in 1985-1988 tournament and Australia hosted the last tournament in 2008.

### *Rugby Union*

Rugby union is one of the region's most prominent sports. Rugby union being the national sport of New Zealand, Samoa, Fiji and Tonga. Fiji's sevens team is one of the most successful in the world, as is New Zealand's. Australia has won the Rugby World Cup a record two times. New Zealand won the inaugural World Cup in 1987. Australia and New Zealand jointly hosted the World Cup in 1987. Australia hosted it in 2003 and New Zealand is to host it in 2011.

### *Cricket*

Cricket is a popular summer sport in Australia and New Zealand. Australia had ruled International cricket as the number one team for more than a decade, and have won the last three Cricket World Cups. New Zealand is also considered a strong competitor in the sport, with the New Zealand Cricket Team, also called the Black Caps, enjoying success in many competitions. Both Australia and New Zealand are full members of the ICC. Fiji, Vanuatu and Papua New Guinea are some of the Associate/Affiliate members of the ICC from Oceania that are governed by ICC East Asia-Pacific. Beach Cricket, a greatly simplified variant of cricket played on a sand beach, is also a popular recreational sport in Australia.

Cricket is culturally a significant sport for summer in Oceania. The Boxing Day Test is very popular in Australia, conducted every year on 26 December at the Melbourne Cricket Ground, Melbourne.

### *Australian Rules Football*

Australian rules football is the national sport in Nauru and is very popular in Australia. It is also very popular in Papua New Guinea.

### *Association Football (Soccer)*

The Oceania Football Confederation (OFC) is one of six association football confederations under the auspices of FIFA, the international governing body of the sport. The OFC is the only confederation without an automatic qualification to the World Cup Finals. Currently the winner of the OFC qualification tournament must play off against an Asian confederation side to qualify for the World Cup..

Currently, Vanuatu is the only country in Oceania to call football its national sport.

Oceania has only been represented at four World Cup Finals — Australia in 1974 and 2006 and New Zealand in 1982 and 2010. However, Australia is now no longer a member of the Oceania Football Confederation, having joined the Asian Football Confederation in 2006.

Both Australia and New Zealand have qualified for the 2010 World Cup making it the first time two countries from Oceania have qualified for the World Cup at the same time, although as mentioned, Australia no longer represents Oceania.

## History and Economy

Important human migrations occurred in the Pacific in prehistoric times, most notably those of the Polynesians from the Asian edge of the ocean to Tahiti and then to Hawaii, New Zealand, and Easter Island.

The ocean was sighted by Europeans early in the 16th century, first by the Spanish explorer Vasco Nunez de Balboa who crossed the Isthmus of Panama in 1513 and named it as *Mar del Sur* (South Sea), and then by the Portuguese explorer Ferdinand Magellan, who sailed the Pacific during his circumnavigation from 1519 to 1522.

In 1564, conquistadors crossed the ocean from Mexico led by Miguel López de Legazpi who sailed to the Philippines and Mariana Islands. For the remainder of the 16th century, Spanish influence was paramount, with ships sailing from Spain to the Philippines, New Guinea, and the Solomon Islands.

During the 17th century, the Dutch, sailing around southern Africa, dominated discovery and trade; Abel Janszoon Tasman discovered Tasmania and New Zealand in 1642. The 18th century marked a burst of exploration by the Russians in Alaska and the Aleutian Islands, the French in Polynesia, and the British in the three voyages of James Cook to the South Pacific and Australia, Hawaii, and the North American Pacific Northwest.

Growing imperialism during the 19th century resulted in the occupation of much of Oceania by European powers, and later, the United States and Japan. Significant contributions to oceanographic knowledge were made by the voyages of HMS *Beagle* in the 1830s, with Charles Darwin aboard; HMS *Challenger* during the 1870s; the USS *Tuscarora* (1873–76); and the German *Gazelle* (1874–76).

Although the United States gained control of the Philippines from Spain in 1898, Japan controlled most of the western Pacific by 1914 and occupied many other islands during World War II. However, by the end of that war, Japan was defeated and the U.S. Pacific Fleet was the virtual master of the ocean. Since the end of World War II, many former colonies in the Pacific have become independent states.

The exploitation of the Pacific's mineral wealth is hampered by the ocean's great depths. In shallow waters of the continental shelves off the coasts of Australia and New Zealand, petroleum and natural gas are extracted, and pearls are harvested along the coasts of Australia, Japan, Papua New Guinea, Nicaragua, Panama, and the Philippines, although in sharply declining volume in some cases.

The Pacific's greatest asset is its fish. The shoreline waters of the continents and the more temperate islands yield herring, salmon, sardines, snapper, swordfish, and tuna, as well as shellfish.

### *Environmental Issues*

Marine pollution is a generic term for the harmful entry into the ocean of chemicals or particles. The biggest culprits are people who use the rivers for disposing of their waste. The rivers then empty into the Ocean, and with it the many chemicals used as fertilizers in agriculture. The excess of oxygen depleting chemicals in the water leads to hypoxia and the creation of a dead zone.

Marine debris, also known as marine litter, is a term used to describe human-created waste that has found itself floating in a lake, sea, ocean or waterway. Oceanic debris tends to accumulate at the centre of gyres and coastlines, frequently washing aground where it is known as beach litter.

## Lands Down Under

The term *Down Under* is a colloquialism referring to Australia. The term comes from the fact that this country is located in the southern hemisphere, below many other countries on the globe.

The persistence of the media use of the term has led to its wide embrace and usage. The Men at Work song "Down Under" became an instantly recognisable, patriotic rallying song. The famous Australian boxing champion Kostya Tszyu was nicknamed "The Thunder From Down Under".

According to Roger Ebert's tongue-in-cheek *Glossary of Movie Terms*, the *Down Under Rule* "No film set in Australia is allowed to use the word Australia in its title where "Down Under" is an acceptable alternative. For example, we don't get *The Rescuers in Australia* or *Quigley in Australia.*"

"Down Under" is also attributed as the inspiration for the underwear brand Gilly Hicks, from the internationally recognised American lifestyle brand and company Abercrombie & Fitch.

In US show Lost, Christian Shephard says that Australia is called "Down Under" because "it's as close as you can get to Hell without getting burned".

Australian Snooker player Neil Robertson currently uses Men At Work's famous hit "Down Under" as his entrance music to a match.

# 7

# Australia

Australia, officially the Commonwealth of Australia, is a country in the Southern Hemisphere comprising the mainland of the Australian continent (the world's smallest), the island of Tasmania and numerous smaller islands in the Indian and Pacific Oceans. Neighbouring countries include Indonesia, East Timor and Papua New Guinea to the north, the Solomon Islands, Vanuatu and New Caledonia to the northeast and New Zealand to the southeast.

For at least 40,000 years before European settlement in the late 18th century, Australia was inhabited by indigenous Australians, who belonged to one or more of the roughly 250 language groups. After sporadic visits by fishermen from the immediate north and discovery by Dutch explorers in 1606, Australia's eastern half was claimed by the British in 1770 and initially settled through penal transportation to the colony of New South Wales, founded on 26 January 1788. The population grew steadily in the following years; the continent was explored and an additional five largely self-governing Crown Colonies were established during the 19th century.

On 1 January 1901, the six colonies became a federation and the Commonwealth of Australia was formed. Since Federation, Australia has maintained a stable liberal democratic political system and is a Commonwealth realm. The population is 22 million, with approximately 60% concentrated in and around the mainland state capitals of Sydney, Melbourne, Brisbane, Perth and Adelaide. The nation's capital city is Canberra, in the Australian Capital Territory.

A prosperous developed country with a multicultural society, Australia ranks highly in many international comparisons of national performance such as human development, quality of life, health care, life expectancy, public education, economic freedom and the protection of civil liberties and political rights. Australia is a member of the United Nations, G20, Commonwealth of Nations, ANZUS, OECD, APEC, Pacific Islands Forum and the World Trade Organization.

## Etymology

Pronounced in Australian English, the name *Australia* is derived from the Latin *australis*, meaning "southern". The country has been referred to colloquially as *Oz* since the early 20th century. *Aussie* is common colloquially, as an adjective and noun for "Australian".

Legends of an "unknown land of the south" (*terra australis incognita*) date back to Roman times and were commonplace in medieval geography, but were not based on any documented knowledge of the continent. The first recorded use of the word *Australia* in English was in 1625, in "A note of Australia del Espiritu Santo, written by Master Hakluyt" and published by Samuel Purchas in *Hakluytus Posthumus*. The Dutch adjectival form *Australische* was used by Dutch East India Company officials in Batavia to refer to the newly discovered land to the south in 1638. *Australia* was used in a 1693 translation of *Les Aventures de Jacques Sadeur dans la Decouverte et le Voyage de la Terre Australe*, a 1676 French novel by Gabriel de Foigny, under the pen-name Jacques Sadeur. Alexander Dalrymple used it in *An Historical Collection of Voyages and Discoveries in the South Pacific Ocean* (1771), referring to the entire South Pacific region. In 1793, George Shaw and Sir James Smith published *Zoology and Botany of New Holland*, in which they wrote of "the vast island, or rather continent, of Australia, Australasia or New Holland". It also appeared on a 1799 chart by James Wilson.

The name *Australia* was popularised by Matthew Flinders, who pushed for the name to be formally adopted as early as 1804. When preparing his manuscript and charts for his 1814 *A Voyage to Terra Australis*, he was persuaded by his patron, Sir Joseph Banks, to use the term *Terra Australis* as this was the name most familiar to the public. Flinders did so, but allowed himself the footnote:

"Had I permitted myself any innovation on the original term, it would have been to convert it to Australia; as being more agreeable to the ear, and an assimilation to the names of the other great portions of the earth."

This is the only occurrence of the word *Australia* in that text; but in Appendix III, Robert Brown's *General remarks, geographical and systematical, on the botany of Terra Australis*, Brown makes use of the adjectival form *Australian* throughout,—the first known use of that form. Despite popular conception, the book was not instrumental in the adoption of the name: the name came gradually to be accepted over the following ten years. Lachlan Macquarie, a Governor of New South Wales, subsequently used the word in his dispatches to England, and on 12 December 1817 recommended to the Colonial Office that it be formally adopted. In 1824, the Admiralty agreed that the continent should be known officially as *Australia*.

## History

Human habitation of Australia is estimated to have begun between 42,000 and 48,000 years ago, possibly with the migration of people by land bridges and short sea-crossings from what is now Southeast Asia. These first inhabitants may have been ancestors of modern Indigenous Australians. At the time of European settlement in the late 18th century, most Indigenous Australians were hunter-gatherers, with a complex oral culture and spiritual values based on reverence for the land and a belief in the Dream time. The Torres Strait Islanders, ethnically Melanesian, were originally horticulturalists and hunter-gatherers.

The first recorded European sighting of the Australian mainland and the first recorded European landfall on the Australian continent were attributed to the Dutch navigator Willem Janszoon, who sighted the coast of Cape York Peninsula on an unknown date in early 1606; he made landfall on 26 February at the Pennefather River on the western shore of Cape York, near the modern town of Weipa. The Dutch charted the whole of the western and northern coastlines of "New Holland" during the 17th century, but made no attempt at settlement. In 1770, James Cook sailed along and mapped the east coast of Australia, which he named New South Wales and claimed for Great Britain. Cook's discoveries prepared the way for establishment of a new penal colony. The British Crown Colony of New South Wales was formed on 26 January 1788, when Captain Arthur Phillip led the First Fleet to Port Jackson. This date became Australia's national day, Australia Day. Van Diemen's Land, now known as Tasmania, was settled in 1803 and became a separate colony in 1825. The United Kingdom formally claimed the western part of Australia in 1828.

Separate colonies were carved from parts of New South Wales: South Australia in 1836, Victoria in 1851, and Queensland in 1859. The Northern Territory was founded in 1911 when it was excised from South Australia. South Australia was founded as a "free province"—it was never a penal colony. Victoria and Western Australia were also founded "free", but later accepted transported convicts. A campaign by the settlers of New South Wales led to the end of convict transportation to that colony; the last convict ship arrived in 1848.

The indigenous population, estimated at 350,000 at the time of European settlement, declined steeply for 150 years following settlement, mainly due to infectious disease. The "Stolen Generations" (removal of Aboriginal children from their families), which historians such as Henry Reynolds have argued could be considered genocide, may have contributed to the decline in the Indigenous population. Such interpretations of Aboriginal history are disputed by conservative commentators such as former Prime Minister John Howard as exaggerated or fabricated for political or ideological reasons. This debate is known within Australia as the History Wars. The Federal government gained the power to make laws with respect to Aborigines following the 1967 referendum. Traditional ownership of land—native title—was not recognised until 1992, when the High Court case *Mabo v Queensland (No 2)* overturned the notion of Australia as *terra nullius* ("land belonging to no one") before European occupation.

A gold rush began in Australia in the early 1850s, and the Eureka Stockade rebellion against mining licence fees in 1854 was an early expression of civil disobedience. Between 1855 and 1890, the six colonies individually gained responsible government, managing most of their own affairs while remaining part of the British Empire. The Colonial Office in London retained control of some matters, notably foreign affairs, defence, and international shipping.

On 1 January 1901, federation of the colonies was achieved after a decade of planning, consultation, and voting. The Commonwealth of Australia was established and it became a dominion of the British Empire in 1907. The Federal Capital Territory (later renamed the Australian Capital Territory) was formed in 1911 as the location for the future federal capital of Canberra. Melbourne was the temporary seat of government from 1901 to 1927 while Canberra was constructed. The Northern Territory was transferred from the control of the South Australian government to the federal parliament in 1911. In 1914, Australia joined Britain in fighting World War I, with support from both the outgoing Liberal Party and the incoming Labour Party. Australians took part in many of the major battles fought on the Western Front. Of about 416,000 who served, about 60,000 were killed and another 152,000 were wounded. Many Australians regard the defeat of the Australian and New Zealand Army Corps (ANZACs) at Gallipoli as the birth of the nation—its first major military action. The Kokoda Track campaign is regarded by many as an analogous nation-defining event during World War II.

Britain's Statute of Westminster 1931 formally ended most of the constitutional links between Australia and the UK. Australia adopted it in 1942, but it was backdated to 1939 to confirm the validity of legislation passed by the Australian Parliament during World War II. The shock of the UK's defeat in Asia in 1942 and the threat of Japanese invasion caused Australia to turn to the United States as a new ally and protector. Since 1951, Australia has been a formal military ally of the US, under the ANZUS treaty. After World War II, Australia encouraged immigration from Europe. Since the 1970s and following the abolition of the White Australia policy, immigration from Asia and elsewhere was also promoted. As a result, Australia's demography, culture, and self-image were transformed. The final constitutional ties between Australia and the UK were severed with the passing of the Australia Act 1986, ending any British role in the government of the Australian

States, and closing the option of judicial appeals to the Privy Council in London. In a 1999 referendum, 55% of Australian voters and a majority in every Australian state rejected a proposal to become a republic with a president appointed by a two-thirds vote in both Houses of the Australian Parliament. Since the election of the Whitlam Government in 1972, there has been an increasing focus in foreign policy on ties with other Pacific Rim nations, while maintaining close ties with Australia's traditional allies and trading partners.

## Politics

Australia is a constitutional monarchy with a federal division of powers. It uses a parliamentary system of government with Queen Elizabeth II at its apex as the Queen of Australia, a role that is distinct from her position as monarch of the other Commonwealth realms. As the Queen resides in the United Kingdom, the executive powers vested in her by the Constitution are normally exercised by her viceroys in Australia (the Governor-General at the federal level and by the Governors at the state level), who by convention act on the advice of her Ministers. The most notable exercise of the Governor-General's reserve powers outside a Prime Minister's request was the dismissal of the Whitlam Government in the constitutional crisis of 1975.

The federal government is separated into three branches:

- The legislature: the bicameral Parliament, comprising the Queen (represented by the Governor-General), the Senate, and the House of Representatives;
- The executive: the Federal Executive Council, in practice the Governor-General as advised by the Prime Minister and Ministers of State;
- The judiciary: the High Court of Australia and other federal courts, whose judges are appointed by the Governor-General on advice of the Council.

In the Senate (the upper house), there are 76 senators: twelve each from the states and two each from the mainland territories (the Australian Capital Territory and the Northern Territory).. The House of Representatives (the lower house) has 150 members elected from single-member electoral divisions, commonly known as "electorates" or "seats", allocated to states on the basis of population, with each original state guaranteed a minimum of five seats. Elections for both chambers are normally held every three years, simultaneously; senators have overlapping six-year terms except for those from the territories, who only have three-year terms; thus only 40 of the 76 places in the Senate are put to each election unless the cycle is interrupted by a double dissolution.

Australia's electoral system uses preferential voting for all lower house elections with the exception of Tasmania and the ACT, which, along with the Senate and most state upper houses, combine it with proportional representation in a system known as the single transferable vote. Voting is compulsory for all enrolled citizens 18 years and over in every jurisdiction, as is enrolment (with the exception of South Australia). Although the Prime Minister is appointed by the Governor-General, in practice the party with majority support in the House of Representatives forms government and its leader becomes Prime Minister.

There are two major political groups that usually form government, federally and in the states: the Australian Labour Party, and the Coalition which is a formal grouping of the Liberal Party and its minor partner, the National Party. Independent members and several minor parties—including the Greens and the Australian Democrats—have achieved representation in Australian parliaments, mostly in upper houses. The last federal election was held in 2007 and brought the Labour Party to office with Kevin Rudd as Prime Minister. Following a partyroom leadership challenge, Julia Gillard became the first female Prime Minister in June 2010.

### *States and Territories*

Australia has six states—New South Wales, Queensland, South Australia, Tasmania, Victoria, and Western Australia—and two major mainland territories—the Northern Territory and the Australian Capital Territory (ACT). In most respects these two territories function as states, but the Commonwealth Parliament can override any legislation of their parliaments. By contrast, federal legislation overrides state legislation only in areas that are set out in Section 51 of the Australian

Constitution; state parliaments retain all residual legislative powers, including those over schools, state police, the state judiciary, roads, public transport, and local government, since these do not fall under the provisions listed in Section 51.

Each state and major mainland territory has its own parliament—unicameral in the Northern Territory, the ACT, and Queensland, and bicameral in the other states. The states are sovereign entities, although subject to certain powers of the Commonwealth as defined by the Constitution. The lower houses are known as the Legislative Assembly (the House of Assembly in South Australia and Tasmania); the upper houses are known as the Legislative Council. The head of the government in each state is the Premier, and in each territory the Chief Minister. The Queen is represented in each state by a Governor; and in the Northern Territory, the Administrator. In the Commonwealth, the Queen's representative is the Governor-General.

The federal parliament directly administers the following territories:

- Jervis Bay Territory, a naval base and sea port for the national capital in land that was formerly part of New South Wales
- Christmas Island and Cocos (Keeling) Islands
- Ashmore and Cartier Islands
- Coral Sea Islands
- Heard Island and McDonald Islands
- Australian Antarctic Territory.

Norfolk Island is also technically an external territory; however, under the Norfolk Island Act 1979 it has been granted more autonomy and is governed locally by its own legislative assembly. The Queen is represented by an Administrator, currently Owen Walsh.

## Foreign Relations and Military

Over recent decades, Australia's foreign relations have been driven by a close association with the United States through the ANZUS pact, and by a desire to develop relationships with Asia and the Pacific, particularly through ASEAN and the Pacific Islands Forum. In 2005 Australia secured an inaugural seat at the East Asia Summit following its accession to the Treaty of Amity and Cooperation in Southeast Asia. Australia is a member of the Commonwealth of Nations, in which the Commonwealth Heads of Government meetings provide the main forum for cooperation.

Australia has pursued the cause of international trade liberalisation. It led the formation of the Cairns Group and Asia-Pacific Economic Cooperation. Australia is a member of the Organisation for Economic Co-operation and Development and the World Trade Organization, and has pursued several major bilateral free trade agreements, most recently the Australia – United States Free Trade Agreement and Closer Economic Relations with New Zealand. As of 2010, Australia is negotiating a free trade agreement with Japan, with whom Australia has close economic ties as a trusted partner in the Asia-Pacific region.

Along with New Zealand, the United Kingdom, Malaysia, and Singapore, Australia is party to the Five Power Defence Arrangements, a regional defence agreement. A founding member country of the United Nations, Australia is strongly committed to multilateralism, and maintains an international aid program under which some 60 countries receive assistance. The 2005–06 budget provides A$2.5 billion for development assistance; as a percentage of GDP, this contribution is less than that recommended in the UN Millennium Development Goals. Australia ranks seventh overall in the Centre for Global Development's 2008 Commitment to Development Index.

Australia's armed forces—the Australian Defence Force (ADF)—comprise the Royal Australian Navy (RAN), the Australian Army, and the Royal Australian Air Force (RAAF), in total numbering 80,561 personnel (including 55,068 regulars and 25,493 reservists). The titular role of Commander-in-Chief is vested in the Governor-General, who appoints a Chief of the Defence Force from one of the armed services on the advice of the government. Day-to-day force operations are under the command of the Chief, while broader administration and the formulation of defence policy is undertaken by the Minister and Department of Defence.

In the 2010–11 budget, defence spending was A$25.7 billion, representing the 14th largest defence budget in the world but accounting for only 1.2% of global military spending. Australia has been involved in UN and regional peacekeeping, disaster relief, and armed conflict; it currently has deployed approximately 3,330 defence force personnel in varying capacities to 12 overseas operations in areas including East Timor, Solomon Islands and Afghanistan.

## Geography of Australia

The Geography of Australia encompasses a wide variety of biogeographic regions being the world's smallest continent but the sixth-largest country in the world. The population of Australia is concentrated along the eastern and southeastern coasts. The geography of the country is extremely diverse, ranging from the snow capped mountains of the Australian Alps and Tasmania to large deserts, tropical and temperate forests.

Neighbouring countries include Indonesia, East Timor and Papua New Guinea to the north, the Solomon Islands, Vanuatu and the French dependency of New Caledonia to the northeast, and New Zealand to the southeast.

### *Physical Geography*

Australia is a country, an island, and a continent. It is located in Oceania between the Indian Ocean and the South Pacific Ocean at 27°S 144°E/27°S 144°E. It is the sixth largest country in the world with a total area of 7,686,850 square kilometres (2,967,909 sq.mi) (including Lord Howe Island and Macquarie Island), making it slightly smaller than the contiguous 48 states of the United States and 31.5 times larger than the United Kingdom.

The Australian mainland has a total coastline length of 35,876 km (22,292 mi) with an additional 23,859 km (14,825 mi) of island coastlines. There are 758 estuaries around the country with most located in the tropical and sub-tropical zones. Australia claims an extensive Exclusive Economic Zone of 8,148,250 square kilometres (3,146,057 sq.mi). This exclusive economic zone does not include the Australian Antarctic Territory. Australia has the largest area of ocean jurisdiction of any country on earth. It has no land borders. The northernmost points of the country are the Cape York Peninsula of Queensland and the Top End of the Northern Territory. The western half of Australia consists of the Western Plateau, which rises to mountain heights near the west coast and falls to lower elevations near the continental centre. The Western Plateau region is generally flat, though broken by various mountain ranges such as the Hamersley Range, the MacDonnell Ranges, and the Musgrave Range. Surface water is generally lacking in the Western Plateau, although there are several larger rivers in the west and north, such as the Murchison, Ashburton, and Victoria river.

The Eastern Highlands, or Great Dividing Range, lie near the eastern coast of Australia, separating the relatively narrow eastern coastal plain from the rest of the continent. These Eastern Australian temperate forests have the greatest relief, the most rainfall, the most abundant and varied flora and fauna, and the densest human settlement.

Between the Eastern Highlands and the Western Plateau lie the Central Lowlands, which are made up of the Great Artesian Basin and Australia's largest river systems, Murray-Darling Basin and Lake Eyre Basin.

Off the eastern coast of Australia is the world's largest coral reef complex, the Great Barrier Reef. The State of Tasmania, a large and mountainous island, resides in the south-eastern corner of Australia.

## *Geology*

Australia is the lowest, flattest, and oldest continental landmass on Earth and it has had a relatively stable geological history. Geological forces such as tectonic uplift of mountain ranges or clashes between tectonic plates occurred mainly in Australia's early history, when it was still a part of Gondwana. Its highest peak is Mount Kosciuszko (2228 m), which is relatively low in comparison to the highest mountains on other continents. Erosion and weathering have heavily weathered Australia's surface.

Australia is situated in the middle of the tectonic plate, and therefore currently has no active volcanism. Minor earthquakes which produce no damage occur regularly, while major earthquakes measuring greater than magnitude 6 occur on average every five years. The terrain is mostly low plateau with deserts, rangelands and a fertile plain in the southeast. Tasmania and the Australian Alps do not contain any permanent icefields or glaciers, although they may have existed in the past. The Great Barrier Reef, by far the world's largest coral reef, lies a short distance off the north-east coast. Mount Augustus, in Western Australia, is the largest monolith in the world.

### *Regions*

The Australian continental landmass consists of 6 distinct landform divisions. These are:

- The Eastern Highlands—including The Great Dividing Range, the fertile Brigalow Belt strip of grassland behind the east coast, and the Eastern Uplands
- The Eastern Alluvial Plains and Lowlands—Murray Darling basin covers southern part, also includes parts of the Lake Eyre Basin and extends to the Gulf of Carpentaria
- The South Australian Highlands—including the Flinders Range, Eyre Peninsula and Yorke Peninsula
- The Western Plateau—including the Nullarbor Plain
- The Central Deserts
- Northern Plateau and Basins—including the Top End.

### *Hydrology*

Because much of Australia's interior is arid, the low average annual rainfall means interior rivers are often dry and lakes empty. The headwaters of some waterways are located in tropical regions where summer rains create a high rate of discharge. Flood events drastically alter the dry environment in which the ecology of central Australia has had to adapt to the boom and bust cycle.

The Great Artesian Basin-an important source of water, it is the world's largest and deepest fresh water basin. Access to water from the basin has led to the expansion of grazing into areas that were previously far too dry for livestock. Towns and cities across the country sometimes face major water storage and usage crisis in which restrictions and other measures are implemented to reduce water consumption. Water restrictions are based on a gradient of activities that become progressively banned as the situation worsens. Billabongs is the Australian name given to oxbow lakes that can form along a meandering river's course. In a world-wide comparison of height, Australia's waterfalls are relatively insignificant, with the longest drop ranked 135th according to the World Waterfall Database.

## Water Security in Australia

Water security in Australia has become a major concern over the course of the late 20th and early 21st century as a result of population growth, severe drought, fears of the effects of global warming on Australia, environmental degradation from reduced environmental flows, competition between competing interests such as grazing, irrigation and urban water supplies, and competition between upstream and downstream users.

Since the turn of the 21st century, there have been attempts to establish a privatised water market in Australia, with the state of Victoria acting as a model for other states. A great deal of political parties, community groups, NGO's and other groups and people see the privatisation of water as a denial of basic human rights on behalf of State and Federal Governments. Although the issue of water privatisation is not commonly understood by the greater Australian population, it remains one of the greatest threats to water security in Australia.

Another major threat to water security in the country is overpopulation, as Australia is a very dry continent overall and the majority of it is uninhabitable and unarable. Australia has an estimated sustainable population of 10 million, a current population of 22 million and a predicted population of 35 million by 2049, this would be more than triple the sustainable limit.

Water reform was first placed on the national agenda at the 1994 Council of Australian Governments (COAG) meeting when a strategic framework was devised.

As the knowledge of surface and groundwater systems grew and the awareness of the significance of sustainable water markets increased, further water reform was agreed to at the 2004 COAG meeting, under a national blueprint known as the National Water Initiative (NWI).

### *Sustainable Water Management*

Efforts to establish sustainable water management plans for the nation and states have repeatedly been overlooked by State and Federal governments instead favouring measures that will appeal to industry and political pressure groups, while selling and marketing messages to the general public that buffer the economic, environmental and social effects of their decisions.

## Competition for Water

### *Competition between States*

In Australia there is competition for the resources of the Darling River system between Queensland, New South Wales and South Australia. Similarly there is competition for the resources of the Murray River between NSW, Victoria and South Australia. The SA government established a new ministerial portfolio of Water Security as the water security issues facing SA and the Lower Lakes and Coorong emerged. SA also has an Independent Commissioner for Water Security.

### *Competition between Regions*

In Victoria, Australia a pipeline from the Goulburn Valley to Melbourne has led to protests by farmers against the linkage of Victoria's water system to facilitate the privatisation of water in the region.

### *Competition between Uses*

In the Macquarie Marshes of NSW grazing and irrigation interests compete for water flowing to the marshes which would otherwise support the environment. There have been a series of embankments built to channel water flowing towards the marshes to privately held commercial interests. Some of these works are thought to have been done illegally.

### *Competition for Environmental Flows*

The Snowy Mountains Scheme diverted water from the Snowy River to the Murray River and the Murrumbidgee River for the benefit of irrigators and electricity generation through hydro-electric power. During recent years government has taken action to increase environmental flows to the Snowy in spite of severe drought in the Murray-Darling Basin. The Australian Government has implemented buy-backs of water allocations, or properties with water allocations, to endeavour to increase environmental flows.

### *National Water Commission*

The National Water Commission is an independent statutory body within the Department of the Environment, Water, Heritage and the Arts that was established under the *National Water Commission Act 2004* to implement the National Water Initiative and reform the broader national water agenda.

The National Water Commission published a report on the future need for desalination technologies to play a role in securing Australia's water supplies. It also publishes a biennial assessment of progress in implementation of the National Water Initiative, the latest being in October 2009. The 2009 assessment nominates areas of slow or inadequate reform and makes 68 recommendations for action over the next two years.

### *Research*

The CSIRO is leading a Water for a Healthy Country Flagship research project to develop information technologies to help water managers.

The Urban Water Security Research Alliance has been formed to address South-East Queensland's emerging urban water issues.

The Wentworth Group of Concerned Scientists conducted research into the sustainability of water usage in Australia.

### *Related Environmental Campaigns*

The WWF campaigns for security of water flows to areas with such as RAMSAR listed wetlands.

### *Snowy River Flows*

The Australian Conservation Foundation and Total Environment Centre campaigned for restoration of some environmental flows to the Snowy River which had been diverted to the Murray River by the Snowy Mountains Scheme for irrigation and power generation. This campaign led to a multi party agreement to restore some flows.

### *National Plan for Water Security*

In January 2005 the Federal Government published "A National Plan For Water Security". This was after the Wentworth Group of Concerned Scientists released its Blueprint for a Living Continent in November, 2002. This blueprint set out a five point plan.

The Inter-governmental Agreement on a National Water Initiative was signed at the 25 June 2004 Council of Australian Governments meeting. The Tasmanian Government joined the Agreement in June 2005 and the Western Australia Government joined in April 2006.

### *Major Capital Works for Increased Major Urban Water Security*

Several major capital works are currently under construction in an effort to privatise the nation's water.

Water desalination has been introduced as a way of creating "new" water to perceivably increase the security of the supply of water, at huge financial costs, environmental losses and increased greenhouse gas emissions, while several pipelines are under construction in an effort to link regional systems to facilitate the trading of water.

The policies of several State Governments moves away from sustainable water management and divests themselves of their responsibilities to provide sustainable, just and affordable, potable water to their population. Privatisation of water in Australia is in turn, of high risk to water security in the country.

### *Western Australia*

The WA government has built the Kwinana Desalination Plant in an effort to privatise Western Australia's water system.

### *New South Wales*

The Kurnell Desalination Plant constructed by the NSW Government at Kurnell near Botany Bay has commenced operations. The power usage is to be offset by a new wind farm. Sydney also draws water from the Shoalhaven River near Nowra.

### *South Australia*

The SA government expects to have a desalination plant operating at Port Stanvac near Lonsdale, South Australia by late 2010

### *Victoria*

The Victorian state government plans several major construction projects to be completed as Private-Public Partnerships (PPPs), in an effort to link state water supplies. These projects are designed to establish a state-wide water market in preparation of the privatisation of Victoria's water. The works include a $3.1 billion desalination plant to be built near Wonthaggi by the end of 2011, a pipeline (the North-South pipeline) to connect water supplies from the Goulburn Valley to Melbourne and Central Victoria, and an inter connector pipeline connecting the Geelong-Ballarat region.

Many argue that in privatising Victoria's water resources and creating a market to trade water between regions, the State Government is divesting itself from its core human rights responsibilities to ensure potable water to the population it represents. This would come at great environmental cost, increase energy usage, decrease efficience, ignore sustainable water management options and increase end-user water costs, placing water privatisation as a high risk to water security in Victoria, despite its acceptance by the state government.

### *Queensland*

The Queensland government has initiated the South East Queensland Water Grid, to enhance the state's water security with the Western Corridor Recycled Water Project being a key part. The Gold Coast desalination plant that has been built at Tugun on the Gold Coast, is another project that will supply "new" water to the future water market in the South East Queensland region. The Queensland government investigated the possibility of building a pipeline from the Northern Rivers of New South Wales to South East Queensland to facilitate water trade between these regions.

### *Murray Darling Basin*

The Snowy Mountains Scheme increased water security to properties near the Murray and Murrumbidgee Rivers during the 1960s and 70s. Over allocation of water licences and prolonged severe drought affecting snowfalls have undermined that security in the early 21st century.

### *Ord River Scheme*

The Ord River Irrigation Scheme (ORIS) in the Kimberley region of Western Australia created Lake Argyle, Australia's largest lake. ORIS provides water for irrigation to over 117 km$^2$ of farmland and there are plans to extend the scheme to allow irrigation of 440 km$^2$ in the future.

### *Lachlan River Valley*

On October 24, 2009 the Sydney Morning Herald reported that the Lachlan River would stop flowing west of Condoblin within weeks as flows are slashed to keep only part of the river flowing, that the Wyangala Dam could be empty by mid summer (January 15, 2010) and that thousands of households will need to have water trucked to them. This situation follows 8 years of drought. Farmers in the area of the Lachlan Valley where the river will not flow are deeply concerned for the viability of their farms. More than 100,000 people live in the total Lachlan catchment. 14% of NSW agricultural production is generated in the region from a land area of approximately 10% of NSW..

### *South Australia*

***The Coorong:*** The Coorong's future is in doubt because of low flows of water reaching the mouth of the Murray River. Acid sulphate soils are being exposed as water levels drop in the Doorong and nearby lakes, adding to the problems in the region.

## Political Geography

Australia consists of six states, two major mainland territories, and other minor territories. The states are New South Wales, Queensland, South Australia, Tasmania, Victoria and Western Australia. The two major mainland territories are the Northern Territory and the Australian Capital Territory.

Australia also has several minor territories; the federal government administers a separate area within New South Wales, the Jervis Bay Territory, as a naval base and sea port for the national capital. In addition Australia has the following, inhabited, external territories: Norfolk Island, Christmas Island, Cocos (Keeling) Islands, and several largely uninhabited external territories: Ashmore and Cartier Islands, Coral Sea Islands, Heard Island and McDonald Islands and the Australian Antarctic Territory.

### *Climate*

By far the largest part of Australia is desert or semi-arid – 40% of the landmass is covered by sand dunes. A total of 18% of Australia's mainland is desert. Only the south-east and south-west corners have a temperate climate and moderately fertile soil. The northern part of the country has a tropical climate: part is tropical rainforests, part grasslands, and part desert.

Rainfall is highly variable, with frequent droughts lasting several seasons thought to be caused in part by the El Nino-Southern Oscillation. Occasionally a dust storm will blanket a region or even several states and there are reports of the occasional large tornado. Rising levels of salinity and desertification in some areas is ravaging the landscape.

Australia's tropical/subtropical location and cold waters off the western coast make most of western Australia a hot desert with aridity, a marked feature of the greater part of the continent. These cold waters produce little moisture needed on the mainland. A 2005 study by Australian and American researchers investigated the desertification of the interior, and suggested that one explanation was related to human settlers who arrived about 50,000 years ago. Regular burning by these settlers could have prevented monsoons from reaching interior Australia. The outback covers 70 percent of the continent.

### *Natural Hazards*

Cyclones along the northern coasts; severe thunderstorms, droughts and occasional floods; frequent bushfires.

### *Environment*

Current issues: soil erosion from over grazing, industrial development, urbanization, and poor farming practices; soil salinity rising due to the use of poor quality water; desertification (partly as a result of the introduction by European settlers of **Rabbits**); introduced pest species; clearing for agricultural purposes threatens the natural habitat of many unique animal and plant species; the Great Barrier Reef off the northeast coast, the largest coral reef in the world, is threatened by increased shipping and its popularity as a tourist site; limited natural fresh water resources; threats from invasive species.

## *International Agreements*

- *Party to:* Antarctic-Environmental Protocol, Antarctic Treaty, Biodiversity, Climate Change, Endangered Species, Environmental Modification, Hazardous Wastes, Law of the Sea, Marine Dumping, Marine Life Conservation, Nuclear Test Ban 1963, Nuclear Non-Proliferation, Ozone Layer Protection, Ship Pollution, Tropical Timber 1994, Wetlands, Whaling
- *Signed, but not ratified:* Climate Change-Kyoto Protocol, Desertification.

## *Environment*

Although most of Australia is semi-arid or desert, it includes a diverse range of habitats from alpine heaths to tropical rainforests, and is recognised as a megadiverse country. Because of the continent's great age, extremely variable weather patterns, and long-term geographic isolation, much of Australia's biota is unique and diverse. About 85% of flowering plants, 84% of mammals, more than 45% of birds, and 89% of in-shore, temperate-zone fish are endemic. Australia has the greatest number of reptiles of any country, with 755 species.

Australian forests are mostly made up of evergreen species, particularly eucalyptus trees in the less arid regions, Wattles replace them in dryer regions and deserts as the most dominant species. Among well-known Australian fauna are the monotremes (the platypus and echidna); a host of marsupials, including the kangaroo, koala, and wombat, and birds such as the emu and the kookaburra. Australia is home to many dangerous animals including some of the most venomous snakes in the world. The dingo was introduced by Austronesian people who traded with Indigenous Australians around 3000 BCE. Many plant and animal species became extinct soon after first human settlement, including the Australian megafauna; others have disappeared since European settlement, among them the thylacine.

Many of Australia's ecoregions, and the species within those regions, are threatened by human activities and introduced plant and animal species. The federal *Environment Protection and Biodiversity Conservation Act 1999* is the legal framework for the protection of threatened species. Numerous protected areas have been created under the National Strategy for the Conservation of Australia's Biological Diversity to protect and preserve unique ecosystems; 65 wetlands are listed under the Ramsar Convention, and 15 natural World Heritage Sites have been established. Australia was ranked 46th of 149 countries in the world on the 2008 Environmental Performance Index.

Climate change has become an increasing concern in Australia in recent years, with many Australians considering protection of the environment to be the most important issue facing the country. The first Rudd Ministry has initiated several emission reduction activities; Rudd's first official act, on his first day in office, was to sign the instrument of ratification of the Kyoto Protocol. Nevertheless, Australia's carbon dioxide emissions per capita are among the highest in the world, lower than those of only a few other industrialised nations. Rainfall in Australia has slightly increased over the past century, both nationwide and for two quadrants of the nation, while annual mean temperatures increased significantly over the past decades. Water restrictions are currently in place in many regions and cities of Australia in response to chronic shortages due to urban population increases and localised drought.

## Economy

The economy of Australia is a developed, modern market economy with a GDP of approximately $1 trillion USD. In 2009, it was the 13th largest national economy by nominal GDP and the 18th largest measured by PPP adjusted GDP, representing about 1.7% of the World economy. Australia was also ranked the 21st largest importer and 23rd largest exporter.

Australia is a member of the APEC, G20, OECD and WTO organisations. Australia has also entered into free trade agreements with ASEAN, Chile, New Zealand, Singapore, Thailand, and the United States. The ANZCERTA agreement with New Zealand has greatly increased integration with the New Zealand economy and there are now plans to form an

Australasian Single Economic Market by 2015. The Australian economy is dominated by its service sector, representing 68% of Australian GDP. The agricultural and mining sectors (10% of GDP combined) account for 57% of the nation's exports.

The Australian dollar is the currency of the Commonwealth of Australia and its territories, including Christmas Island, Cocos (Keeling) Islands, and Norfolk Island. It is also the official currency of the independent Pacific Island nations of Kiribati, Nauru and Tuvalu.

The Australian Securities Exchange is the largest stock exchange in Australia.

## *Overview*

Australia's per-capita GDP is slightly higher than that of the UK, Germany, and France in terms of purchasing power parity. The country was ranked second in the United Nations 2009 Human Development Index and sixth in *The Economist* worldwide quality-of-life index 2005.

The emphasis on exporting commodities rather than manufactures has underpinned a significant increase in Australia's terms of trade during the rise in commodity prices since 2000. Australia's current account is more than 7% of GDP negative: Australia has had persistently large current account deficits for more than 50 years. Australia has grown at an average annual rate of 3.6% for over 15 years, well above the OECD average of 2.5%. Australia's average GDP growth rate for the period 1901-2000 is at 3.4% annually.

As of December 2009, there were approximately 10,844,000 people employed, with an unemployment rate of 5.5%. Over the past decade, inflation has typically been 2–3% and the base interest rate 5–6%. The service sector of the economy, including tourism, education and financial services, constitutes 69% of GDP.

Rich in natural resources, Australia is a major exporter of agricultural products, particularly wheat and wool, minerals such as iron-ore and gold, and energy in the forms of liquified natural gas and coal. Although Agriculture and natural resources constitute only 3% and 5% of GDP, respectively, they contribute substantially to export performance. Australia's largest export markets are Japan, China, South Korea, India and the USA.

In the past decade, one of the most significant sectoral trends experienced by the economy has been the growth (in relative terms) of the mining sector (including petroleum). In terms of contribution to GDP, this sector grew from around 4.5% in 1993-94, to almost 8% in 2006-07.

Growth in the services sector has also grown considerably, with property and business services in particular growing from 10% to 14.5% of GDP over the same period, making it the largest single component of GDP (in sectoral terms). This growth has largely been at the expense of the manufacturing sector, which in 2006-07 accounted for around 12% of GDP. A decade earlier, it was the largest sector in the economy, accounting for just over 15% of GDP.

### *Economic Liberalisation*

From the early 1980s onwards, the Australian economy has undergone a continuing economic liberalisation. In 1983, under Prime Minister Bob Hawke, but mainly driven by Treasurer Paul Keating, the Australian dollar was floated and financial deregulation was undertaken.

In 2000, the introduction of a goods and services tax (GST) sought to encourage the level of saving amongst lower income earners. To combat the consequential reduction in consumption for low income earners, income taxes were lowered as a trade-off for the introduction of the GST. The overall level of taxation in Australia has since been consistently reduced to encourage private consumption and investment, as opposed to higher government expenditure.

### *Current Areas of Concern*

Current areas of concern to some economists include Australia's large current account deficit, Australia's current account deficit for the 2007-2008 financial year was up 4% to $19.49 billion (according to the Australian Bureau of Statistics), the absence of a successful export-oriented manufacturing industry, an Australian property bubble, and high levels of net foreign debt owed by the private sector. Professor Steve Keen has written extensively about consumer/household indebtedness and the level of home prices relative to income. While there is some talk about increasing levels of government debt as a result of increased federal government deficit caused by falling tax revenues and stimulus during 2008 and proposed to continue through 2010, Australia has one of the lowest levels of government debt as a proportion of GDP among developed countries, largely as a result of the debt repayment program pursued by then Treasurer Peter Costello and the then Howard government. The coal, electricity generation and agricultural industries have concerns about the impact of requirements to reduce carbon emissions under the proposed Carbon Pollution Reduction Scheme — Which, partly in response, has been shelved until 2012. The price of housing in terms of median incomes has been highlighted by a recent Demographia survey with Australian capital city residential housing being among the most expensive in the world. A long drought and its impacts on retail food costs and export volumes of crops and meat and the possible impacts of climate change on agriculture has also been of concern.

## Taxation

Taxation in Australia is levied at the federal, state and local government levels. Taxes vary from state to state due to their different needs, populations, economics and budgetary positions.

The Commonwealth raises revenue from personal income taxes and business taxes. Other taxes include the goods and services tax (GST), excise and customs duties. The Commonwealth is the main source of income for state governments. As a result of state dependence on federal taxation revenue to meet decentralised expenditure responsibilities, Australia is said to suffer from a vertical fiscal imbalance.

### *State Taxation*

Besides receipts of funds from the Commonwealth, states and territories also have their own taxes to enable them to fund the services they provide. The types and tax rates vary from state/territory to state/territory. State taxes commonly include payroll tax levied on businesses, a poker machine tax levied on businesses who offer gambling services, land tax levied on people and businesses who own land and most significantly, stamp duty levied on sales of land (in every state) and other items (chattels in some states, unlisted shares in others, and even sales of contracts in some states).

### *Federal-state Financial Arrangements*

The states effectively lost the ability to raise income tax during the Second World War. In 1942, the Commonwealth invoked its Constitutional taxation power and enacted the *Income Tax Act* and three other statutes to levy a uniform income tax across the Commonwealth. These Acts sought to raise the funds necessary to meet burgeoning wartime expenses and reduce the unequal tax burden between the states by replacing state income taxes with a centralised tax system. The legislation could not expressly prohibit state income taxes does not curtail the power of states to levy taxes) but the Commonwealth's proposal made localised income tax both practically impossible and politically poisonous. The Commonwealth offered instead compensatory grants authorised by s.96 of the Constitution for the loss of state income.

The states rejected the Commonwealth's regime and challenged the legislation's validity in the First Uniform Tax Case (*South Australia v Commonwealth*) of 1942. The High Court of Australia held that each of the statutes establishing Commonwealth income tax was a valid use of the s.51(ii) power, with Latham C J noting that the system did not undermine essential state functions and imposed only economic and political pressure upon them.

The Second Uniform Tax Case (*Victoria v Commonwealth* (1957)) reaffirmed the Court's earlier decision and confirmed the power of the Commonwealth to make s.96 grants conditionally (in this case, a grant made on the condition that the recipient state does not levy income tax).

Since the Second Uniform Tax Case, a number of other political and legal decisions have centralised fiscal power with the Commonwealth. In *Ha vs. New South Wales* (1997), the High Court found that the *Business Franchise Licences (Tobacco) Act 1987* (NSW) was invalid because it levied a customs duty, a power exercisable only by the Commonwealth (s.90). This decision effectively invalidated state taxes on cigarettes, alcohol and petrol. Similarly, the imposition of a Commonwealth goods and services tax (GST) in 2000 transferred another revenue base to the Commonwealth.

Consequently, Australia has one of the most pronounced vertical fiscal imbalances in the world, with states collecting just 18% of all governmental revenues but responsible for almost 50% of the spending and policy areas. Furthermore, the centralisation of revenue collection has allowed the Commonwealth to force state policy in areas well beyond the scope of its Constitutional powers by using the grants power (s.96) to mandate the terms on which the states spend money in areas over which the Commonwealth has no power (such as spending on education, health and policing).

### *Municipal Taxation*

Local governments (called councils in Australia) have their own taxes (called rates) to enable them to provide rubbish collection, park maintenance services, libraries and museums, etc.

### *Trade and Economic Performance*

In the second half of the twentieth century, Australian trade shifted away from Europe and North America to Japan and other East Asian markets. Regional franchising businesses, now a $128 billion sector, have been operating co-branded sites overseas for years with new investors coming from Western Australia and Queensland.

The Australian economy has been performing nominally better than other economies of the OECD and has supported economic growth for 16 consecutive years. According to the Reserve Bank of Australia, Australian per capita GDP growth is higher than that of New Zealand, US, Canada and The Netherlands. The past performance of the Australian economy has been heavily influenced by US, Japanese and Chinese economic growth. Despite high global demand for Australian mineral commodities, export growth has remained flat in comparison to strong import growth. Even though Australia enjoys high commodity prices, economists have warned that structural change is needed in order to increase the size of manufacturing sector.

### *Chinese Investment*

There is substantial export to China of iron ore, wool, and other raw materials and over 120,000 Chinese students study in Australian schools and universities. China is a major purchaser of Australian debt. In 2009, offers were made by state-owned Chinese companies to invest 22 billion dollars in Australia's resource extraction industry.

### *Australia's Balance of Payments*

In trade terms, the Australian economy has had persistently large current account deficits for more than 50 years. One single factor that undermines balance of payments is Australia's narrow export base.

Dependent upon commodities, the Australian governments has endeavoured to redevelop the Australian manufacturing sector. This initiative, also known as microeconomic reform, has helped Australian manufacturing to grow from 10.1% in 1983-1984 to 17.8% in 2003-2004.

There are other factors that have contributed to the extremely high current account deficit that Australia has today. Lack of international competitiveness and heavy reliance on capital goods from overseas might increase Australia's current account deficit in the future.

The Australian dollar is the currency for the nation, including Christmas Island, Cocos (Keeling) Islands, and Norfolk Island, as well as the independent Pacific Island states of Kiribati, Nauru, and Tuvalu. After the 2006 merger of the Australian Stock Exchange and the Sydney Futures Exchange, the Australian Securities Exchange is now the ninth largest in the world.

Ranked third in the Index of Economic Freedom (2010), Australia's per capita GDP is slightly higher than that of the United States, United Kingdom, Germany, and France. The country was ranked second in the United Nations 2009 Human Development Index, first in Legatum's 2008 Prosperity Index, and sixth in *The Economist* worldwide Quality-of-Life Index for 2005. All of Australia's major cities fare well in global comparative liveability surveys; Melbourne reached second place on *The Economist*'s 2008 World's Most Livable Cities list, followed by Perth, Adelaide, and Sydney in fourth, seventh and ninth place respectively.

An emphasis on exporting commodities rather than manufactured goods has underpinned a significant increase in Australia's terms of trade since the start of the century, due to rising commodity prices. Australia has a balance of payments that is more than 7% of GDP negative, and has had persistently large current account deficits for more than 50 years. Australia has grown at an average annual rate of 3.6% for over 15 years, in comparison to the OECD annual average of 2.5%. Australia was one of the few OECD nations to avoid falling into a technical recession during the late 2000s global financial downturn.

The Hawke Government floated the Australian dollar in 1983 and partially deregulated the financial system. The Howard Government followed with a partial deregulation of the labour market and the further privatisation of state-owned businesses, most notably in the telecommunications industry. The indirect tax system was substantially changed in July 2000 with the introduction of a 10% Goods and Services Tax (GST). In Australia's tax system, personal and company income tax are the main sources of government revenue.

In January 2007, there were 10,033,480 people employed, with an unemployment rate of 4.6%. Over the past decade, inflation has typically been 2–3% and the base interest rate 5–6%. The service sector of the economy, including tourism, education, and financial services, accounts for 69% of GDP. Although agriculture and natural resources account for only 3% and 5% of GDP respectively, they contribute substantially to export performance. Australia's largest export markets are Japan, China, the US, South Korea, and New Zealand.

### *Demography*

Most of the estimated 22 million Australians are descended from colonial-era settlers and post-Federation immigrants from Europe, with almost 90% of the population being of European descent. For generations, the vast majority of immigrants came from the British Isles, and the people of Australia are still mainly of British or Irish ethnic origin. In the 2006 Australian census, the most commonly nominated ancestry was Australian (37.13%), followed by English (31.65%), Irish (9.08%), Scottish

(7.56%), Italian (4.29%), German (4.09%), Chinese (3.37%), and Greek (1.84%).

Australia's population has quadrupled since the end of World War I, spurred by an ambitious immigration program. Following World War II and through to 2000, almost 5.9 million of the total population settled in the country as new immigrants, meaning that nearly two out of every seven Australians were born overseas. Most immigrants are skilled, but the immigration quota includes categories for family members and refugees. The Federal Government estimates that cutting immigration from 280,000 to its target of 180,000 will result in a population of 36 million by 2050.

In 2001, 23.1% of Australians were born overseas; the five largest immigrant groups were those from the United Kingdom, New Zealand, Italy, Vietnam, and China. Following the abolition of the White Australia policy in 1973, numerous government initiatives have been established to encourage and promote racial harmony based on a policy of multiculturalism. In 2005–06, more than 131,000 people emigrated to Australia, mainly from Asia and Oceania. The migration target for 2006–07 was 144,000. The total immigration quota for 2008–09 is around 300,000—its highest level since the Immigration Department was created after World War II.

The Indigenous population—mainland Aborigines and Torres Strait Islanders—was counted at 410,003 (2.2% of the total population) in 2001, a significant increase from 115,953 in the 1976 census. A large number of Indigenous people are not identified in the Census due to undercount and cases where their Indigenous status is not recorded on the form; after adjusting for these factors, the ABS estimated the true figure for 2001 to be around 460,140 (2.4% of the total population).

Indigenous Australians experience higher than average rates of imprisonment and unemployment, lower levels of education, and life expectancies for males and females that are 11–17 years lower than those of non-indigenous Australians. Some remote Indigenous communities have been described as having "failed state"-like conditions.

In common with many other developed countries, Australia is experiencing a demographic shift towards an older population, with more retirees and fewer people of working age. In 2004, the average age of the civilian population was 38.8 years. A large number of Australians (759,849 for the period 2002–03) live outside their home country.

## *Language*

English is the national language. Australian English has a unique accent and a small number of unique terms, some of which have found their way into other dialects of the English-speaking world. It has less internal dialectal variation than either British or American English. Grammar and spelling are largely based on those of British English. According to the 2006 census, English is the only language spoken in the home for close to 79% of the population. The next most common languages spoken at home are Italian (1.6%), Greek (1.3%) and Cantonese (1.2%); a considerable proportion of first-and second-generation migrants are bilingual.

Between 200 and 300 indigenous Australian languages are thought to have existed at the time of first European contact, of which only about 70 have survived. Many of these are exclusively spoken by older people; only 18 indigenous languages are still spoken by all age groups. At the time of the 2006 Census, 52,000 indigenous Australians, representing 12% of the indigenous population, reported that they spoke an indigenous language at home. Australia has a sign language known as Auslan, which is the main language of about 5,500 deaf people.

## *Religion*

Australia has no state religion. In the 2006 census, 64% of Australians listed themselves as Christian, including 26% as Roman Catholic and 19% as Anglican. About 19% of the population cited "No religion" (which includes humanism, atheism, agnosticism, and rationalism), which was the fastest-growing group from 2001 to 2006, and a further 12% did not answer (the question is optional) or did not give a response adequate for interpretation. The second-largest religion in Australia is Buddhism (2.1%), followed by Islam (1.7%), Hinduism (0.8%), and Judaism (0.5%). Overall, fewer than 6% of Australians identify with non-Christian religions. Weekly attendance at church services in 2004 was about 1.5 million:

about 7.5% of the population. Religion does not play a central role in the lives of much of the population.

### *Education*

School attendance is compulsory throughout Australia. All children receive 11 years of compulsory education from the age of 6 to 16 (Year 1 to 10), before they can undertake two more years (Years 11 and 12), contributing to an adult literacy rate that is assumed to be 99%. A preparatory year prior to Year 1, although not compulsory, is almost universally undertaken. In the Programme for International Student Assessment, Australia regularly scores among the top five of thirty major developed countries (member countries of the Organisation for Economic Co-operation and Development). Government grants have supported the establishment of Australia's 38 universities; all but one is public. There is a state-based system of vocational training, known as TAFE Institutes, and many trades conduct apprenticeships for training new tradespeople. Approximately 58% of Australians aged from 25 to 64 have vocational or tertiary qualifications, and the tertiary graduation rate of 49% is the highest among OECD countries. The ratio of international to local students in tertiary education in Australia is the highest in the OECD countries.

### *Health*

Life expectancy in Australia is relatively high, with figures of 78.7 years for boys and 83.5 years for girls born in 2006. Australia has the highest rates of skin cancer in the world, while cigarette smoking is the largest preventable cause of death and disease. Australia has one of the highest proportions of overweight citizens amongst developed nations; it has also been one of the most successful in managing the spread of HIV/AIDS.

Australia introduced universal health care, known as Medibank, in 1975. Reworked by successive governments, its current incarnation, Medicare came into existence in 1984. It is now nominally funded by an income tax surcharge known as the *Medicare levy*, currently set at 1.5%. Traditionally, the management of public health has been split between the state and federal governments. The states manage hospitals and attached outpatient services, while the commonwealth funds the Pharmaceutical Benefits Scheme (reducing the costs of medicines) and general practice. Under the Rudd government, a health reform plan has emerged which will allow the federal government to take "full responsibility for primary health care", essentially taking control of hospitals and outpatient services from the states. Total expenditure on health (including private sector spending) is around 9.8 per cent of GDP.

## Culture

Since 1788, the primary basis of Australian culture has been Anglo-Celtic Western culture. Distinctive cultural features have also arisen from Australia's natural environment and Indigenous cultures. Since the mid-20th century, American popular culture has strongly influenced Australia, particularly through television and cinema. Other cultural influences come from neighbouring Asian countries, and through large-scale immigration from non-English-speaking nations.

### *Art*

Australian visual arts are thought to have begun with the cave and bark paintings of its Indigenous peoples. The traditions of Indigenous Australians are largely transmitted orally and are tied to ceremony and the telling Dream time stories. From the time of European settlement, a theme in Australian art has been the natural landscape, seen for example in the works of Albert Namatjira, Arthur Streeton and others associated with the Heidelberg School, and Arthur Boyd.

The country's landscape remains a source of inspiration for Australian modernist artists; it has been depicted in acclaimed works by the likes of Sidney Nolan, Fred Williams, Sydney Long, and Clifton Pugh. Australian artists influenced by modern American and European art include cubist Grace Crowley, surrealist James Gleeson, abstract expressionist Brett Whiteley, and pop artist Martin Sharp. Contemporary Indigenous Australian art is the only art movement of international significance to emerge from Australia and "the last great art movement of the 20th century"; its exponents have included Emily Kngwarreye. Art critic Robert Hughes has written several influential books about Australian history and art, and was described as the "world's most famous art critic" by *The New*

*York Times*. The National Gallery of Australia and state galleries maintain Australian and overseas collections.

### *Performing Arts*

Many of Australia's performing arts companies receive funding through the federal government's Australia Council. There is a symphony orchestra in each state, and a national opera company, Opera Australia, well-known for its famous soprano Joan Sutherland. At the start of the 20th century, Nellie Melba was one of the world's leading opera singers. Ballet and dance are represented by The Australian Ballet and various state companies. Each state has a publicly funded theatre company.

### *Media*

The Australian cinema industry began with the 1906 release of *The Story of the Kelly Gang*, which is regarded as being the world's first feature-length film, but both Australian feature film production and the distribution of British-made features declined dramatically after World War I as American studios and distributors monopolised the industry and by the 1930s around 95% of the feature films screened in Australia were produced in Hollywood. By the late 1950s feature film production in Australia had effectively ceased and there were no all-Australian feature films made in the decade between 1959 and 1969.

Thanks to initiatives by the Gorton and Whitlam federal governments, the New Wave of Australian cinema of the 1970s brought provocative and successful films, some exploring the nation's colonial past, such as *Picnic at Hanging Rock* and *Breaker Morant*, while the so-called "Ocker" genre produced several highly successful urban-based comedy features including *The Adventures of Barry Mackenzie* and *Alvin Purple*. Later hits included *Mad Max* and *Gallipoli*. More recent successes included *Shine* and *Rabbit-Proof Fence*. Notable Australian actors include Judith Anderson, Errol Flynn, Nicole Kidman, Hugh Jackman, Heath Ledger, Geoffrey Rush, and current joint director of the Sydney Theatre Company, Cate Blanchett.

Australian literature has also been influenced by the landscape; the works of writers such as Banjo Paterson, Henry

Lawson, and Dorothea Mackellar captured the experience of the Australian bush. The character of the nation's colonial past, as represented in early literature, is popular with modern Australians. In 1973, Patrick White was awarded the Nobel Prize in Literature, the first Australian to have achieved this. Australian winners of the Man Booker Prize have included Peter Carey, Thomas Keneally and Iris Murdoch; David Williamson, and David Malouf are also renowned writers, and Les Murray is regarded as "one of the leading poets of his generation".

Australia has two public broadcasters (the Australian Broadcasting Corporation and the multicultural Special Broadcasting Service), three commercial television networks, several pay-TV services, and numerous public, non-profit television and radio stations. Each major city has at least one daily newspaper, and there are two national daily newspapers, *The Australian* and *The Australian Financial Review*. In 2008, Reporters Without Borders placed Australia 25th on a list of 173 countries ranked by press freedom, behind New Zealand (7th) and the United Kingdom (23rd) but ahead of the United States (48th). This low ranking is primarily because of the limited diversity of commercial media ownership in Australia; most print media are under the control of News Corporation and Fairfax Media.

## *Cuisine*

Australia's food traditions have been largely shaped by its inhabitants. For most of Australian history, the cuisine was based on traditional Indigenous bushfood using native berries, fruit, fish, kangaroo and even insects such as the witchetty grub. Later, British food was introduced by the first European settlers; the Sunday roast has become an enduring tradition for many Australians. In the 19th and especially the 20th century, food was increasingly influenced by Mediterranean and Asian cultures of immigrants of the period.

## *Sport*

Around 24% Australians over the age of 15 regularly participate in organised sporting activities in Australia. Australia has strong international teams in cricket, field hockey, netball, rugby league and rugby union, having been Olympic or world champions at least twice in each sport in the last 25 years for both men and women where applicable. Australia is also powerful in track cycling, rowing, and swimming, having consistently been in the top-five medal-winners at Olympic or World Championship level since 2000. Swimming is the strongest of these sports; Australia is the second-most prolific medal winner in the sport in Olympic history.

Some of Australia's most successful sportspersons are swimmers Dawn Fraser, Murray Rose, Shane Gould and Ian Thorpe; sprinter Betty Cuthbert; tennis players Rod Laver, Ken Rosewall, Evonne Goolagong, and Margaret Court; cricketer Donald Bradman; three-time Formula One world champion Jack Brabham; five-times motorcycle grand prix world champion Mick Doohan; and prodigious billiards player Wally Lindrum. Nationally, other popular sports include Australian rules football, horse racing, surfing, football (soccer), and motor racing.

Australia has participated in every summer Olympics of the modern era, and every Commonwealth Games. Australia hosted the 1956 Summer Olympics in Melbourne and the 2000 Summer Olympics in Sydney, and has ranked among the top six medal-takers since 2000. Australia has also hosted the 1938, 1962, 1982, and 2006 Commonwealth Games. Other major international events held in Australia include the Australian Open tennis grand slam tournament, international cricket matches, and the Australian Formula One Grand Prix. The highest-rating television programs include sports telecasts such as the summer Olympics, State of Origin, and the grand finals of the National Rugby League and Australian Football League.

## Polynesia

Polynesia is a subregion of Oceania, comprising a large grouping of over 1,000 islands scattered over the central and southern Pacific Ocean. The people who inhabit the islands of Polynesia are termed Polynesians and they share many similar traits including language, culture and beliefs.

The term "Polynesia" was first used by Charles de Brosses, a French writer, in 1756, and originally applied to all the islands of the Pacific. In 1831, Jules Dumont d'Urville proposed a restriction on its use during a lecture to the Geographical Society of Paris.

### *Geographic Area*

Polynesia is generally defined as the islands within the Polynesian Triangle although this does not cover the geographic spread and settlement of Polynesian people across a greater area. Geographically, and oversimply, the Polynesian Triangle may be seen with its corners at Hawaii, New Zealand and Easter Island.

The other main island groups located within the Polynesian triangle are Samoa, Tonga, the Cook Islands, Tuvalu, Tokelau, Niue, Wallis and Futuna and French Polynesia.

A Polynesian island group outside of this great triangle is Rotuma situated north of the Fijian islands. There are also small outlier Polynesian enclaves in Papua New Guinea, the Solomons, The Caroline Islands, some of the Lau group to Fiji's southeast and in Vanuatu.

However, in essence, Polynesia is an anthropological term referring to one of the three parts of Oceania (the others being Micronesia and Melanesia) whose pre-colonial population generally belongs to one ethno-cultural family as a result of centuries of maritime migrations.

### *Island Groups*

The following are the islands and island groups, either nations or subnational territories, that are of native Polynesian culture. Some islands of Polynesian origin are outside the general triangle that geographically defines the region.

### *Main Polynesia*

- American Samoa (overseas United States territory)
- Cook Islands (self-governing state in free association with New Zealand)
- Easter Island (called *rapa nui* in Rapa Nui, politically part of Chile)
- French Polynesia ("overseas territory", a territory of France)
- Hawaii (a state of the United States)
- New Zealand (independent nation)
- Niue (self-governing state in free association with New Zealand)

- Rotuma (in Fiji)
- Samoa (independent nation)
- Tokelau (overseas dependency of New Zealand)
- Tonga (independent nation)
- Tuvalu (independent nation)
- Wallis and Futuna (overseas territory of France).

***Polynesian Outliers***

- Norfolk Island (an Australian External Territory)
- Pitcairn Islands (a British Overseas Territory).

***In Melanesia***

- Anuta (in the Solomon Islands)
- Mele (in Vanuatu)
- Bellona Island (in the Solomon Islands)
- Emae (in Vanuatu)
- Nuguria (in Papua New Guinea)
- Nukumanu (in Papua New Guinea)
- Ontong Java (in the Solomon Islands)
- Pileni (in the Solomon Islands)
- Rennell (in the Solomon Islands)
- Sikaiana (in the Solomon Islands)
- Takuu (in Papua New Guinea)
- Tikopia (in the Solomon Islands).

***In Micronesia***

- Kapingamarangi (in the Federated States of Micronesia)
- Nukuoro (in the Federated States of Micronesia).

## History of the Polynesian People

### *Mainstream Theories*

The Polynesian people are considered to be by ancestry a subset of the sea-migrating Austronesian people and the tracing of Polynesian languages places their prehistoric origins in the Malay archipelago. There are three theories regarding the spread of humans across the Pacific to Polynesia. These are outlined well by Kayser et al., (2000) and are as follows:

- Express Train model: A recent (c. 3,000 years ago) expansion out of Southeast Asia, predominantly Taiwan, via Melanesia but with little genetic admixture between those migrating and the existing native population, reaching western Polynesian islands around 2,000 years ago. This theory is supported by the majority of current genetic, linguistic, and archaeological data.
- Entangled Bank model: Supposes a long history of cultural and genetic interactions amongst southeast Asians, Melanesians, and already-established Polynesians.
- Slow Boat model: Similar to the express-train model but with a longer hiatus in Melanesia along with admixture, both genetically, culturally and linguistically with the local population. This is supported by the Y-chromosome data of Kayser et al. [2000], which shows that all three haplotypes of Polynesian Y chromosomes can be traced back to Melanesia

Between about 3000 and 1000 BC speakers of Austronesian languages spread throughout the islands of Southeast Asia. These people, according to linguistic and archaeological evidence, originated from aborigines in Taiwan as tribes whose natives were thought to have arrived through South China about 8,000 years ago to the edges of western Micronesia and on into Melanesia, although they are different from the Han Chinese who now form the majority of people in China and Taiwan. In fact Taiwan, previously inhabited mostly by non-Han aborigines, was Sinicized via large-scale migration accompanied with assimilation during the 1600s.

In the archaeological record there are well-defined traces of this expansion which allow the path it took to be followed and dated with a degree of certainty. It is thought that roughly 3,500 years BP the Lapita culture appeared in the Bismarck Archipelago, northwest Melanesia. This culture is argued to have either been developed there or, more likely, to have spread from China/Taiwan. The most eastern site for Lapita archaeological remains recovered so far through archaeology in Samoa is at Mulifanua on Upolu. The Mulifanua site, where 4,288 pottery sherds have been found and studied, has a true age of circa 3,000 BP based on C14 dating.

Within a mere three or four centuries between about 1300 and 900 BC, the Lapita culture spread 6,000 km further to the east from the Bismarck Archipelago, until it reached as far as Fiji, Tonga, and Samoa which were populated around 2,000 years ago. In this region, the distinctive Polynesian culture developed.

The spread of pottery and domesticates in Polynesia are connected with the Lapita culture that started expanding from New Guinea to as far east as Fiji, Samoa, and Tonga. During this time the aspects of the Polynesian culture developed. Around 300 BC this new Polynesian people spread eastward from Fiji, Samoa, and Tonga to the Cook Islands, Tahiti, the Tuamotus, and the Marquesas Islands. This was supported by Patrick Kirch and Marshall Weisler when they performed X-ray fluorescence sourcing of basalt artifacts found on both islands.

Between AD 300 and 500, the Polynesians discovered and settled Easter Island. This is supported by archaeological evidence as well as the introduction of flora and fauna consistent with the Polynesian culture and characteristic of the tropics to this subtropical island. However, more recently, Jared Diamond and others have proposed the date of arrival to be closer to 800 AD, as scientific uncertainly clouded the evidence of an earlier arrival. Furthermore, Terry Hunt of the University of Hawaii has proposed that the arrival did not take place until approximately 1200 AD. Around AD 500, Hawai'i was settled by the Polynesians and around AD 1000 New Zealand was settled as well.

## *Polynesian Links to the Americas*

The sweet potato *kumara*, which is native to the Americas, was widespread in Polynesia when Europeans first reached the Pacific. *Kumara* has been radiocarbon-dated in the Cook Islands to 1000 AD, and current thinking is that it was brought to central Polynesia circa 700 CE and spread across Polynesia from there, possibly by Polynesians who had travelled to South America and back.

Thor Heyerdahl proposed in the mid-twentieth century that the Polynesians had migrated from South America on balsa-log boats.

## *Cultures of Polynesia*

Polynesia divides into two distinct cultural groups, East Polynesia and West Polynesia. The culture of West Polynesia is conditioned to high populations. It has strong institutions of marriage and well-developed judicial, monetary and trading traditions. It comprises the groups of Tonga, Niue, Samoa and the northwestern Polynesian outliers.

Eastern Polynesian cultures are highly adapted to smaller islands and atolls, principally the Cook Islands, Tahiti, the Tuamotus, the Marquesas, Hawaii, Rapa Nui and smaller central-pacific groups. The large islands of New Zealand were first settled by Eastern Polynesians who adapted their culture to a non-tropical environment.

Unlike in Melanesia, leaders were chosen in Polynesia based on their hereditary bloodline. Samoa however, had another system of government that combines elements of heredity and real-world skills to choose leaders. This system is called Fa'amatai.

Religion, farming, fishing, weather prediction, out-rigger canoe (similar to modern catamarans) construction and navigation were highly developed skills because the population of an entire island depended on them. Trading of both luxuries and mundane items was important to all groups. Many low-lying islands could suffer severe famine if their gardens were poisoned by the salt from the storm-surge of a hurricane. In these cases fishing, the primary source of protein, would not ease loss of food energy. Navigators, in particular, were highly respected and each island maintained a house of navigation with a canoe-building area.

Settlements by the Polynesians were of two categories: the hamlet and the village. Size of the island inhabited determined whether or a not a hamlet would be built. The larger volcanic islands usually had hamlets because of the many zones that could be divided across the island. Food and resources were more plentiful and so these settlements of four to five houses (usually with gardens) were established so that there would be no overlap between the zones. Villages, on the other hand, were built on the coasts of smaller islands and consisted of thirty or more houses—in the case of atolls, on only one of the group so that food cultivation was on the others. Usually these villages were fortified with walls and palisades made of stone and wood.

However, New Zealand demonstrates the opposite: large volcanic islands with fortified villages.

As well as being great navigators these people were artists and artisans of great skill. Simple objects, such as fish-hooks would be manufactured to exacting standards for different catches and decorated even when the decoration was not part of the function. In some island groups weaving was a strong part of the culture and gifting woven articles an ingrained practice. Stone and wooden weapons were considered to be more powerful the better they were made and decorated. Dwellings were imbued with character by the skill of their building. Body decoration and jewellery is of international standard to this day.

The religious attributes of Polynesians were common over the whole Pacific region. While there are some differences in their spoken languages they largely have the same explanation for the creation of the earth and sky, for the gods that rule aspects of life and for the religious practices of everyday life. People travelled thousands of miles to celebrations that they all owned communally.

Due to relatively large numbers of competitive sects of Christian missionaries in the islands, many Polynesian groups have been converted to Christianity. Polynesian languages are all members of the family of Oceanic languages, a sub-branch of the Austronesian language family.

### *Polynesian Languages*

One basic language prevails throughout Polynesia. The vowels are consistently the same-a, e, i, o, and u, pronounced as in Italian or Spanish-and the consonants are always followed by a vowel. Dialects have developed in various island groups by changes in consonant sounds. *R* and *v* are used in central and eastern Polynesia whereas *l* and *v* are used in western Polynesia. In some dialects certain consonants are not fully sounded but are represented by an inverted comma (or 'okina) over its place in the word. In the Society Islands, *k* and *ng* were dropped; so the name for the ancestral homeland, common throughout Polynesia and pronounced Havaiki in other dialects of central Polynesia, is here pronounced Havai'i. In New Zealand, where *w* is used instead of *v*, the ancient home is Hawaiki. In the Cook Islands, where *h* is dropped, it is 'Avaiki. In the Hawaiian islands, where *w* is used and *k* is dropped, the largest island of the group is named Hawai'i. In Samoa, where *s* replaces *h*, *v* is preferred to *w*, and *k* is dropped in formal Samoan, the largest island is called Savai'i.

### *Economy*

With the exception of New Zealand, the majority of independent Polynesian islands derive much of their income from foreign aid and remittances from those who live in other countries. Some encourage their young people to go where they can earn good money to remit to their stay-at-home relatives. Many Polynesian locations, such as Easter Island, supplement this with tourism income. Some have more unusual sources of income, such as Tuvalu which marketed its '.tv' internet top-level domain name or the Cooks that relied on stamp sales.

### *Polynesian Navigation*

Polynesia comprised islands diffused throughout a triangular area with sides of four thousand miles. The area from the Hawaiian Islands in the north, to Easter Island in the east and to New Zealand in the south were all settled by Polynesians.

Navigators travelled to small inhabited islands using only their own senses and knowledge passed by oral tradition from navigator to apprentice. In order to locate directions at various times of day and year, navigators in Eastern Polynesia memorized important facts: the motion of specific stars, and where they would rise on the horizon of the ocean; weather; times of travel; wildlife species (which congregate at particular positions); directions of swells on the ocean, and how the crew would feel their motion; colours of the sea and sky, especially how clouds would cluster at the locations of some islands; and angles for approaching harbours.

These way finding techniques along with outrigger canoe construction methods, were kept as guild secrets. Generally each island maintained a guild of navigators who had very high status; in times of famine or difficulty these navigators could trade for aid or evacuate people to neighbouring islands. To this day, original traditional methods of Polynesian Navigation are still taught in the Polynesian outlier of Taumako Island in the Solomon Islands.

From a single chicken bone recovered from the archaeological site of El Arenal-1, on the Arauco Peninsula, Chile, a 2007 research report looking at radiocarbon dating and an ancient DNA sequence indicate that Polynesian navigators may have reached the Americas at least 100 years before Columbus (who arrived 1492 AD), introducing chickens to South America. A later report looking at the same specimens concluded:

A published, apparently pre-Columbian, Chilean specimen and six pre-European Polynesian specimens also cluster with the same European/Indian subcontinental/Southeast Asian sequences, providing no support for a Polynesian introduction of chickens to South America. In contrast, sequences from two archaeological sites on Easter Island group with an uncommon haplogroup from Indonesia, Japan, and China and may represent a genetic signature of an early Polynesian dispersal. Modeling of the potential marine carbon contribution to the Chilean archaeological specimen casts further doubt on claims for pre-Columbian chickens, and definitive proof will require further analyses of ancient DNA sequences and radiocarbon and stable isotope data from archaeological excavations within both Chile and Polynesia.

Knowledge of the traditional Polynesian methods of navigation were largely lost after contact with and colonization by Europeans. This left the problem of accounting for the presence of the Polynesians in such isolated and scattered parts of the Pacific. By the late 19th century to the early 20th century a more generous view of Polynesian navigation had come into favour, perhaps creating a romantic picture of their canoes, seamanship and navigational expertise.

In the mid to late 1960s, scholars began testing sailing and paddling experiments related to Polynesian navigation: David Lewis sailed his catamaran from Tahiti to New Zealand using stellar navigation without instruments and Ben Finney built a 40-foot replica of a Hawaiian double canoe "Nalehia" and tested it in Hawaii. Meanwhile, Micronesian ethnographic research in the Caroline Islands revealed that traditional stellar navigational methods were still in every day use. Recent re-creations of Polynesian voyaging have used methods based largely on Micronesian methods and the teachings of a Micronesian navigator, Mau Piailug.

It is probable that the Polynesian navigators employed a whole range of techniques including use of the stars, the movement of ocean currents and wave patterns, the air and sea interference patterns caused by islands and atolls, the flight of birds, the winds and the weather. Scientists think that long-distance Polynesian voyaging followed the seasonal paths of birds. There are some references in their oral traditions to the flight of birds and some say that there were range marks onshore pointing to distant islands in line with these flyways. One theory is that they would have taken a frigatebird with them. These birds refuse to land on the water as their feathers will become waterlogged making it impossible to fly. When the voyagers thought they were close to land they may have released the bird, which would either fly towards land or else return to the canoe. It is likely that the Polynesians also used wave and swell formations to navigate. It is thought that the Polynesian navigators may have measured the time it took to sail between islands in "canoe-days" or a similar type of expression.

Also, people of the Marshall Islands used special devices called stick charts, showing the places and directions of swells and wave-breaks, with tiny seashells affixed to them to mark the positions of islands along the way. Materials for these maps were readily available on beaches, and their making was simple; however, their effective use needed years and years of study.

# 8

# New Zealand

New Zealand is an island country in the south-western Pacific Ocean comprising two main landmasses (the North Island and the South Island), and numerous smaller islands, most notably Stewart Island/Rakiura and the Chatham Islands. The indigenous Maori language name for New Zealand is Aotearoa, commonly translated as *The Land of the Long White Cloud*. The Realm of New Zealand also includes the Cook Islands and Niue (self-governing but in free association); Tokelau; and the Ross Dependency (New Zealand's territorial claim in Antarctica).

New Zealand is notable for its geographic isolation: it is situated about 2,000 kilometres (1,200 mi) southeast of Australia across the Tasman Sea, and its closest neighbours to the north are New Caledonia, Fiji and Tonga. During its long isolation New Zealand developed a distinctive fauna dominated by birds, a number of which became extinct after the arrival of humans and the mammals they introduced.

The majority of New Zealand's population is of European descent; the indigenous Maori are the largest minority. Asians and non-Maori Polynesians are also significant minority groups, especially in urban areas. The most commonly spoken language is English.

New Zealand is a developed country that ranks highly in international comparisons on many topics, including education, economic freedom, and lack of corruption. Its cities also consistently rank among the world's most liveable. Elizabeth II, as the Queen of New Zealand, is the country's head of state and is represented by a Governor-General, and executive political power is exercised by the Cabinet of New Zealand.

## Etymology

It is unknown whether Maori had a name for New Zealand as a whole before the arrival of Europeans, although they referred to the North Island as Te Ika a Maui (the fish of Maui) and the South Island as Te Wai Pounamu (the waters of greenstone) or Te Waka o Aoraki (the canoe of Aoraki). Until the early 20th century, the North Island was also referred to as Aotearoa (colloquially translated "land of the long white cloud"); in modern Maori usage, this name refers to the whole country. Aotearoa is also commonly used in this sense in New Zealand English, where it is sometimes used alone, and in some formal uses combined with the English name to express respect to the original inhabitants of the country, for example in the form of "[Organisation name] of Aotearoa New Zealand".

The first European name for New Zealand was *Staten Landt*, the name given to it by the Dutch explorer Abel Tasman, who in 1642 became the first European to see the islands. Tasman assumed it was part of a southern continent connected with land discovered in 1615 off the southern tip of South America by Jacob Le Maire, which had been named *Staten Landt*, meaning "Land of the (Dutch) States-General".

The name *New Zealand* originated with Dutch cartographers, who called the islands *Nova Zeelandia*, after the Dutch province of Zeeland. No one is certain exactly who first coined the term, but it first appeared in 1645 and may have been the choice of cartographer Johan Blaeu. British explorer James Cook subsequently anglicised the name to New Zealand. There is no connection to the Danish island Zealand.

Although the North and South Islands have been known by these names for many years, the New Zealand Geographic Board has stated that as of 2009, they have no official names. The board intends to make these their official names, along with alternative Maori names. Some early maps refer to what is currently known as the South Island as the *Middle Island.* Although several Maori names have been used, Maori Language Commissioner Erima Henare sees Te Ika-a-Maui and Te Wai Pounamu respectively as the most likely choices.

## History

### *Polynesian Settlers*

New Zealand is one of the most recently settled major landmasses. The first known settlers were Eastern Polynesians who, according to most researchers, arrived by canoe in about AD 1250–1300. Some researchers have suggested an earlier wave of arrivals dating to as early as AD 50–150; these people then either died out or left the islands. Over the following centuries these settlers developed into a distinct culture now known as Maori. The population was divided into iwi (tribes) and hapu (subtribes) which would cooperate, compete and sometimes fight with each other. At some point a group of Maori migrated to the Chatham Islands where they developed their distinct Moriori culture.

### *European Explorers*

The first Europeans known to have reached New Zealand were Dutch explorer Abel Janszoon Tasman and his crew in 1642. Maori killed several of the crew and no Europeans returned to New Zealand until British explorer James Cook's voyage of 1768–71. Cook reached New Zealand in 1769 and mapped almost the entire coastline. Following Cook, New Zealand was visited by numerous European and North American whaling, sealing and trading ships. They traded European food and goods, especially metal tools and weapons, for Maori timber, food, artefacts and water. On occasion, Europeans traded goods for sex.

The potato and the musket transformed Maori agriculture and warfare, although the resulting Musket Wars died out once the tribal imbalance of arms had been rectified. From the early nineteenth century, Christian missionaries began to settle New Zealand, eventually converting most of the Maori population, although their initial inroads were mainly among the more disaffected elements of society.

Becoming aware of the lawless nature of European settlement and of increasing French interest in the territory, the British government appointed James Busby as British Resident to New Zealand in 1832. Busby failed to bring law and order to European settlement, but did oversee the introduction of the first national flag on 20 March 1834, after an unregistered New Zealand ship was seized in Australia. The nebulous United Tribes of New Zealand later, in October 1835, sent the Declaration of the Independence of New Zealand to King William IV of the United Kingdom, asking him for protection. Ongoing unrest and the legal standing of the Declaration of Independence prompted the Colonial Office to send Captain William Hobson RN to New Zealand to claim sovereignty for the British Crown and negotiate a treaty with the Maori. The Treaty of Waitangi was first signed in the Bay of Islands on 6 February 1840. The drafting was done hastily and confusion and disagreement continue to surround the translation. The Treaty however remains regarded as New Zealand's foundation as a nation and is revered by Maori as a guarantee of their rights.

In response to New Zealand Company attempts to establish a separate colony in Wellington, and French claims in Akaroa, Hobson, now Lieutenant-Governor, declared British sovereignty over all of New Zealand on 21 May 1840. The two proclamations published in the *New Zealand Advertiser and Bay of Islands Gazette* issue of 19 June 1840 "assert[s] on the grounds of Discovery, the Sovereign Rights of Her Majesty over the Southern Islands of New Zealand, commonly called 'The Middle Island' (South Island) and 'Stewart's Island' (Stewart Island/ Rakiura); and the Island, commonly called 'The Northern Island', having been ceded Sovereignty to Her Majesty." The second proclamation expanded on how sovereignty over the "Northern Island" had been ceded under the treaty signed that February.

Under British rule, New Zealand had initially been part of the colony of New South Wales, but became a separate Crown colony in 1841. Hobson initially selected Okiato as the capital in 1840, before moving the seat of government to Auckland in 1841. Increasing numbers of European settlers came to New Zealand particularly from the British Isles. The Maori were initially eager to trade with the 'Pakeha', as they called them, and many iwi became wealthy. As settler numbers increased, conflicts over land led to the New Zealand Land Wars of the 1860s and 1870s, resulting in the loss of much Maori land. The details of European settlement and the acquisition of land from Maori remain controversial. Representative government for the colony was provided for in 1852 when the United Kingdom passed the New Zealand Constitution Act 1852. The 1st New Zealand Parliament met in 1854. In 1856 the colony became effectively self-governing with the grant of responsible government over all domestic matters other than native policy. Power in this respect would be transferred to the colonial administration in the 1860s.

In 1863 Premier Alfred Domett moved a resolution that the capital transfer to a locality in Cook Strait, apparently due to concern that the South Island might form a separate colony. Commissioners from Australia (chosen for their neutral status) advised that Wellington was suitable because of its harbour and central location, and parliament officially sat there for the first time in 1865. In 1893 the country became the first nation in the world to grant all women the right to vote.

## *20th Century*

In 1907 New Zealand declared itself a Dominion within the British Empire. In 1947 the country adopted the Statute of Westminster, making New Zealand a Commonwealth realm, although in practice Britain had long since ceased to play a significant role in governing New Zealand. As the country became more politically independent, however, it became more dependent economically; in the 1890s, refrigerated shipping allowed the export of meat and dairy products to Britain, a trade which provided the basis for strong economic growth in New Zealand.

New Zealand was an enthusiastic member of the British Empire, fighting in the Boer War, World War I and World War II, especially in the Battle of Britain, and supporting Britain in the Suez Crisis. The country was very much a part of the world economy and suffered as others did in the Great Depression of the 1930s. The depression led to the election of the first Labour government, which established a comprehensive welfare state and a protectionist economy.

New Zealand experienced increasing prosperity following World War II. However, some social problems were developing; Maori had begun to leave traditional rural life and move to the cities in search of work. A Maori protest movement eventually developed, which criticised Eurocentrism and worked for more recognition of Maori culture and the Treaty of Waitangi, which they felt had not been fully honoured. In 1975, a Waitangi Tribunal was set up to investigate alleged breaches of the Treaty, and it was enabled to investigate historic grievances in 1985. In common with other developed countries, social developments accelerated in the 1970s and social and political mores changed.

Britain's membership of the European Economic Community in 1973 drastically reduced access for New Zealand exporters to their previous largest market. In the 1953 two-thirds of New Zealand's exports went to Britain, by 2003 this figure had reduced to 4.65%. This and the oil shocks of the 1970s led to significant economic and social changes during the 1980s under the 4th Labour government largely led by Finance Minister Roger Douglas, whose policies are commonly referred to as *Rogernomics*.

## Politics

### *Government*

New Zealand is a constitutional monarchy with a parliamentary democracy. Although it has no codified constitution, the Constitution Act 1986 is the principal formal statement of New Zealand's constitutional structure. The constitution has been described as "largely unwritten" and a "mixture of statutes and constitutional convention." Queen Elizabeth II is the head of state and is titled Queen of New Zealand under the Royal Titles Act 1974. She is represented by the Governor-General, whom she appoints on the exclusive advice of the Prime Minister. The current Governor-General is Anand Satyanand.

The Governor-General exercises the Crown's prerogative powers, such as the power to appoint and dismiss ministers and to dissolve Parliament, and in rare situations, the reserve powers. The Governor-General also chairs the Executive Council, which is a formal committee consisting of all ministers of the Crown. The main *constitutional* function of the governor-general is to "arrange for the leader of the majority political party to form a government"; by constitutional convention, the governor-general "acts on the advice of ministers who have majority support in parliament." Legislative power is held by the democratically elected Parliament of New Zealand, and the rest of the cabinet. The Queen and Governor-General cannot normally exercise any power without the advice of the Cabinet, except in circumstances where their is no cabinet or cabinet has lost the confidence of Parliament.

Members of the Executive Council are required to be Members of Parliament, and most are also in cabinet. Cabinet is the most senior policy-making body and is led by the Prime Minister, who is also, by convention, the Parliamentary leader of the governing party or coalition. This is the highest policymaking body in the government.

The New Zealand Parliament comprises one chamber, the House of Representatives, which usually seats 120 members.

Parliamentary general elections are held every three years under a form of proportional representation called Mixed Member Proportional. The Economist magazine explains:

Under MMP (Mixed Member Proportional) there is usually a 120-seat parliament; an extra seat can sometimes be added to ensure truly proportional representation. Of the total number of seats, 65 electorate (directly elected constituency) seats are contested on the old first-past-the-post basis, including seven seats reserved for the indigenous Maori people. The remaining 55 or so seats are allocated so that representation in parliament reflects overall support for each party (the party vote).

Under the MMP system, a party has either to win a constituency seat or more than 5% of the total party vote in order to gain representation in parliament. The government can continue to rule only if it retains majority support in the House of Representatives, or can secure the support of other political parties to give it a majority to pass legislation and survive parliamentary confidence votes.

The 2008 general election created an 'overhang' of two extra seats, occupied by the Maori Party, due to that party winning more seats in electorates than the number of seats its proportion of the party vote would have given it.

From October 2005 until November 2008, the Labour-led government was in formal coalition with the Progressive Party, Jim Anderton being its only MP. In addition, New Zealand First and United Future provided confidence and supply in return for their leaders being ministers outside cabinet. An arrangement was also made with the Green Party, which gave a commitment not to vote against the government on confidence and supply. In 2007 Labour also had the proxy vote of Taito Phillip Field, a former Labour MP. These arrangements assured the government of a majority of seven MPs on confidence votes.

Labour was defeated by the National Party in the general elections of 8 November 2008. Following the victory, National leader John Key moved quickly to form a government, negotiating coalition agreements with the right-wing ACT party, led by Rodney Hide, the centrist United Future party, albeit with its single seat held by leader Peter Dunne, and the Maori Party, led by Tariana Turia and Pita Sharples. Each of these leaders hold ministerial posts but remain outside of Cabinet. There are three parties in Opposition: the Labour Party, led by Phil Goff; the Greens, co-led by Metiria Turei and Russel Norman and the Progressive Party, under Jim Anderton.

The highest court in New Zealand is the Supreme Court of New Zealand, established in 2004 following the passage of the Supreme Court Act 2003. The act abolished the option to appeal to the Privy Council in London. The current Chief Justice is Dame Sian Elias. New Zealand's judiciary also includes the Court of Appeal; the High Court, which deals with serious criminal offences and civil matters at the trial level and with appeals from lower courts and tribunals; and subordinate courts.

While the Judiciary can sometimes place limits on acts of Parliament, and the 1990 New Zealand Bill of Rights enables some review by the Judiciary of executive action, there is no document ascertaining formal power of judicial review. Its constitutional independence from Parliament is maintained by non-political appointments and strict rules regarding tenure in office. New Zealand is the only country in the world in which all the highest offices in the land have been occupied simultaneously by women: (Queen) Elizabeth II, (Governor-General) Dame Silvia Cartwright, (Prime Minister) Helen Clark, (Speaker of the New Zealand House of Representatives) Margaret Wilson and (Chief Justice) Dame Sian Elias were all in office between March 2005 and August 2006. New Zealand's largest listed company, Telecom New Zealand, had a woman– Theresa Gattung – as its CEO at the time.

### Foreign Relations and the Military

New Zealand maintains a strong profile on environmental protection, human rights and free trade, particularly in agriculture. New Zealand is a member of Commonwealth of Nations, OECD, Five Powers Defence Arrangements, APEC, East Asia Summit, and the United Nations. New Zealand is party to a number of free trade agreements, of which the most important are the New Zealand – China Free Trade Agreement and Closer Economic Relations with Australia.

For its first hundred years, New Zealand followed the United Kingdom's lead on foreign policy. In declaring war on Germany on 3 September 1939, Prime Minister Michael Savage proclaimed, "Where she goes, we go; where she stands, we stand."

The two world wars had a marked impact, with New Zealand losing many young men in places like Gallipoli (where the ANZAC tradition was formed with Australia), Crete, El Alamein and Cassino. New Zealand also played a key part with Britain in the two famous battles, the naval Battle of the River Plate and the Battle of Britain fought in the air. During the Pacific part of World War II, the United States had more than 400,000 American military personnel stationed in New Zealand to prepare for crucial battles such as Tarawa, Guadalcanal, Saipan and Iwo Jima.

After the war the United States exerted an increased influence on culture and the New Zealand people gained a clearer sense of national identity. New Zealand joined Australia and the United States in the ANZUS security treaty in 1951, and later fought alongside the United States in both the Korean and the Vietnam Wars. In contrast, the United Kingdom became increasingly focused on its European interests following the Suez Crisis, and New Zealand was forced to develop new markets after the UK joined the EEC in 1973.

New Zealand has traditionally worked closely with Australia, whose foreign policy followed a similar historical trend. This close bond was formed in Gallipoli and is part of the *ANZAC spirit*, which forms a cornerstone in both countries. In turn, many Pacific Islands such as Western Samoa have looked to New Zealand's lead. The American influence on New Zealand was weakened by the disillusionment with the Vietnam War, the sinking of the Rainbow Warrior by France (which Britain and the US failed to criticise), and by disagreements over environmental and agricultural trade issues and New Zealand's nuclear-free policy.

While the ANZUS treaty was once fully mutual between Australia, New Zealand and the United States, this is no longer the case. In February 1985, New Zealand refused nuclear-powered or-armed ships access to its ports. New Zealand became a Nuclear-free zone in June 1987, the first Western-allied state to do so. In 1986, the United States announced that it was suspending its treaty security obligations to New Zealand pending the restoration of port access.

The New Zealand Nuclear Free Zone, Disarmament, and Arms Control Act 1987 prohibits the stationing of nuclear weapons on the territory of New Zealand and the entry into New Zealand waters of nuclear armed or propelled ships. This legislation remains a source of contention and the basis for the United States' continued suspension of treaty obligations to New Zealand.

Within New Zealand, there have been various wars between iwi, and between the British settlers and iwi. New Zealand has fought in the Second Boer War, World War I, World War II, the Korean War, the Malayan Emergency (and committed troops, fighters and bombers to the subsequent confrontation with Indonesia), the Vietnam War, the Gulf War and the Afghanistan War.

It has also sent a unit of army engineers to help rebuild Iraqi infrastructure for one year during the Iraq War. As of October 2009, New Zealand forces were still active in Afghanistan.

The New Zealand Defence Force has three branches: the New Zealand Army, the Royal New Zealand Navy and the Royal New Zealand Air Force. New Zealand considers its own national defence needs to be modest; it dismantled its air combat capability in 2001. New Zealand has contributed forces to recent regional and global peacekeeping missions, including those in Cyprus, Somalia, Bosnia and Herzegovina, the Sinai, Angola, Cambodia, the Iran/Iraq border, Bougainville, East Timor, and the Solomon Islands.

### *Local Government and External Territories*

The early European settlers divided New Zealand into provinces. These were abolished in 1876 so that government could be centralised, for financial reasons. As a result, New Zealand has no separately represented subnational entities such as provinces, states or territories, apart from local government. However the spirit of the provinces lives on, and there is fierce rivalry exhibited in sporting and cultural events. Since 1876, local government has administered the various regions of New Zealand.

In 1989, the government completely reorganised local government, implementing the current two-tier structure of regional councils and territorial authorities constituted under the Local Government Act 2002. The Resource Management Act 1991 replaced the Town and Country Planning Act as the main planning legislation for local government.

New Zealand has 12 regional councils for the administration of regional environmental and transport matters and 73 territorial authorities that administer roading, sewerage, building consents, and other local matters. The territorial authorities are 16 city councils, 57 district councils, and the Chatham Islands Council.

Four of the territorial councils (one city and three districts) and the Chatham Islands Council also perform the functions of a regional council and are known as unitary authorities. Territorial authority districts are not subdivisions of regional council districts, and a few of them straddle regional council boundaries.

The regions are (asterisks denote unitary authorities): Northland, Auckland, Waikato, Bay of Plenty, Gisborne*, Hawke's Bay, Taranaki, Manawatu-Wanganui, Wellington, Marlborough*, Nelson*, Tasman*, West Coast, Canterbury, Otago, Southland, Chatham Islands*.

As a major South Pacific nation, New Zealand has a close working relationship with manv Pacific Island nations, and continues a political association with the Cook Islands, Niue and Tokelau. New Zealand operates Scott Base in its Antarctic territory, the Ross Dependency. Other countries also use Christchurch to support their Antarctic bases and the city is sometimes known as the "Gateway to Antarctica".

## Geography and Environment

New Zealand comprises two main islands, the North and South Islands, *Te Ika a Maui* and *Te Wai Pounamu* respectively in Maori, and a number of smaller islands, located near the centre of the water hemisphere. The North and South Islands are separated by Cook Strait, 20 kilometres wide at its narrowest point. The total land area, 268,021 square kilometres (103,483 sq mi), is a little less than that of Italy or Japan, and a little more than the United Kingdom.

The country extends more than 1,600 kilometres (990 mi) along its main, north-north-east axis, with approximately 15,134 km (9,404 mi) of coastline.

The most significant of the smaller inhabited islands include Stewart Island/Rakiura; Waiheke Island, in Auckland's Hauraki Gulf; Great Barrier Island, east of the Hauraki Gulf; and the Chatham Islands, named *Rckohu* by Moriori. The country has extensive marine resources, with the seventh-largest Exclusive Economic Zone in the world, covering over four million square kilometres (1.5 million square miles), more than 15 times its land area.

The dramatic and varied landscape of New Zealand has made it a popular location for the production of television programmes and films, including the *Lord of the Rings* trilogy and *The Last Samurai*. The South Island is the largest land mass of New Zealand, and is divided along its length by the Southern Alps. There are 18 peaks over 3,000 metres (9,800 ft) the highest of which is Aoraki/Mount Cook at 3,754 metres (12,316 ft). The top of South Island contains areas of forest in the Kahurangi and other national parks. The south-western corner of South Island is Fiordland, an area of high mountains cut through with steep fjords.

The North Island is less mountainous but is marked by volcanism. The highly active Taupo volcanic zone has formed a large volcanic plateau, the North Island's highest mountain, Mount Ruapehu 2,797 metres (9,177 ft), and a deep caldera filled by Lake Taupo, the country's largest lake. The island's north is a flatter area, once covered by huge kauri trees.

The country owes its varied topography, and perhaps even its emergence above the waves, to the dynamic boundary it straddles between the Pacific and Indo-Australian Plates. New Zealand is part of Zealandia, a microcontinent nearly half the size of Australia that gradually submerged after breaking away from the Gondwanan supercontinent.

About 25 million years ago, a shift in plate tectonic movements began to contort and crumple the region. This is now most evident in the Southern Alps, formed by compression of the crust beside the Alpine Fault. Elsewhere the plate boundary involves the subduction of one plate under the other, producing the Puysegur Trench to the south, the Hikurangi Trench east of the North Island, and the Kermadec and Tonga Trenches further north.

The latitude of New Zealand, from approximately 34 to 47° S, corresponds closely to that of Italy in the Northern Hemisphere. However, its isolation from continental influences and exposure to cold southerly winds and ocean currents give the climate a much milder character. The climate throughout the country is mild and temperate, mainly maritime, with temperatures rarely falling below 0 °C (32 °F) or rising above 30 °C (86 °F) in populated areas. Historical maxima and minima are 42.4 °C (108.3 °F) in Rangiora, Canterbury and "21.6 °C ("6.9 °F) in Ophir, Otago.

Conditions vary sharply across regions from extremely wet on the West Coast of the South Island to semi-arid (Koppen BSh) in the Mackenzie Basin of inland Canterbury and subtropical in Northland. Of the main cities, Christchurch is the driest, receiving only 640 millimetres (25 in) of rain per year; Auckland, the wettest, receives almost twice that amount. Auckland, Wellington and Christchurch all receive a yearly average in excess of 2000 hours of sunshine. The southern and south-western parts of the South Island have a cooler and cloudier climate, with around 1400–1600 hours; the northern and north-eastern parts of the South Island are the sunniest areas of the country and receive approximately 2400–2500 hours.

### *Biodiversity*

Because of its long isolation from the rest of the world and its island biogeography, New Zealand has extraordinary flora and fauna, descended from Gondwanan wildlife or since arriving by flying, swimming or being carried across the sea. About 80% of New Zealand's flora is endemic, including 65 endemic genera. The two main types of forest are those dominated by podocarps and/or the giant kauri, and in cooler climates the southern beech. The remaining vegetation types in New Zealand are grasslands of tussock and other grasses, usually in sub-alpine areas, and the low shrublands between grasslands and forests.

Until the arrival of humans, 80% of the land was forested. A diverse range of megafauna inhabited the forests, including the flightless moas (now extinct), four species of kiwi, the kakapo and the takahc, all endangered by human actions. Unique birds capable of flight included the Haast's eagle, which was the world's largest bird of prey (now extinct), and the large kaka and kea parrots.

Reptiles present in New Zealand include skinks, geckos and the living fossil tuatara. There are four endemic species of primitive frogs. There are no snakes and there is only one venomous spider, the katipo, which is rare and restricted to coastal regions.

There are many endemic species of insect, including the weta, one species of which may grow as large as a house mouse and is the heaviest insect in the world. It was long thought that New Zealand never had any non-marine native mammals, barring three species of bat (one now extinct). However in 2006 scientists discovered bones over 15 million years old from a unique, mouse-sized land mammal in the Otago region of the South Island.

New Zealand's native wildlife has suffered a high rate of extinctions, including around fifty bird species such as the moa, huia, laughing owl, adzebills, and flightless wrens (which occupied the roles elsewhere occupied by mice). This is due to human activities such as hunting, and pressure from introduced feral animals, such as weasels, stoats, cats, goats, deer and brushtailed possums. Five indigenous vascular plant species are now believed to be extinct, including Adam's mistletoe and a species of forget-me-not. Several species have been rediscovered after being thought extinct, with the takahe being the largest.

New Zealand has led the world in island restoration projects, where offshore islands are cleared of introduced mammalian pests and native species are reintroduced. Several islands, including two of the Chatham Islands, are wildlife reserves where common pests such as possums and rodents have been eradicated to allow the reintroduction of endangered species to the islands. A more recent development is the analogous mainland ecological island. Active management has helped increase the population of certain species dramatically. For instance, only five Black Robins remained in 1980, including just one fertile female. There are now around 250, all descended from that one bird.

### *Economy*

New Zealand has a modern, prosperous, developed economy with an estimated GDP (PPP) of US$119.549 billion (2010). The country has a relatively high standard of living with an estimated GDP per capita of US$31,067 in 2010, comparable to that of Southern Europe. Since 2000 New Zealand has made substantial gains in median household income. During the financial crisis of 2007–2010 GDP shrank for five consecutive quarters, the longest recession in over thirty years.

New Zealanders have a high level of life satisfaction as measured by international surveys; this is despite lower GDP per-head levels than many other OECD countries. The country was ranked 20th on the 2009 Human Development Index and 15th in *The Economist's* 2005 worldwide quality-of-life index. The country was ranked 1st in social capital and 10th in overall prosperity in the 2009 Legatum Institute prosperity index. In addition, the 2009 Mercer Quality of Living Survey ranked Auckland 4th place and Wellington 12th place in the world on its list. Taxation in New Zealand is lighter than in other OECD countries. New Zealand a relatively laissez-faire capitalist economy according to the Fraser Institute think tank.

The service sector is the largest sector in the economy (68.8% of GDP), followed by manufacturing and construction (26.9% of GDP) and the farming/raw materials extraction (4.3% of GDP).

New Zealand is heavily dependent on free trade, particularly in agricultural products. Exports account for around 24% of its output, which is a relatively high figure (it is around 50% for many smaller European countries). This makes New Zealand particularly vulnerable to international commodity prices and global economic slowdowns. Its principal export industries are agriculture, horticulture, fishing and forestry. These make up about half of the country's exports. Its major export partners are Australia 20.5%, US 13.1%, Japan 10.3%, China 5.4%, UK 4.9% (2006). Tourism plays a significant role in New Zealand's economy. In 2010 the sector contributed $15.0 billion (or 9.1%) to New Zealand's total GDP and supported 184,800 full-time equivalent jobs (9.6% of the total workforce in New Zealand). Tourists to New Zealand are expected to increase at a rate of 2.5% annually up to 2015.

The New Zealand dollar is the currency of New Zealand. It also circulates in the Cook Islands, Niue, Tokelau, and the Pitcairn Islands. It is sometimes informally known as the "Kiwi dollar".

*The Economist* magazine's outlook for New Zealand (2009) foresees the government's fiscal position to remain tenuous because of "weak revenue growth and rising expenditure". Government debt is expected to balloon from 25% (2008) to 40% (2013). GDP growth will contract in 2009 by 2.6%, then average 2.2% from 2010 to 2013 (although there are "downside risks" which may hamper this growth).

Government will continue to pursue foreign trade. Inflation will be 1.4% in 2009, 1.3% in 2010 and average 2.3% from 2011 to 2013. The New Zealand dollar is expected to weaken against the dollar through 2010, but begin strengthening again beginning 2011 (but the report notes that exchange rates are volatile and hard to predict). The study doesn't see a change of government from the centre-right National Party until possibly 2011.

### *Recent History*

Historically New Zealand enjoyed a high standard of living which relied on its strong relationship with the United Kingdom, and the resulting stable market for its commodity exports. New Zealand's economy was also built upon on a narrow range of primary products, such as wool, meat and dairy products. High demand for these products created sustained periods of economic prosperity, such as the New Zealand wool boom of 1951. However, in 1973 the United Kingdom joined the European Community which effectively ended this particularly close economic relationship between the two countries.

During the 1970s other factors such as the oil crises (1973 and 1979) undermined the viability of the New Zealand economy; which for periods before 1973 had achieved levels of living standards exceeding both Australia and Western Europe.

These events led to a protracted and very severe economic crisis, during which living standards in New Zealand fell behind those of Australia and Western Europe, and by 1982 New Zealand was the lowest in per-capita income of all the developed nations surveyed by the World Bank.

Since 1984, successive governments have engaged in major macroeconomic restructuring, transforming New Zealand from a highly protectionist and regulated economy to a liberalised free-trade economy. These changes are commonly known as Rogernomics and Ruthanasia after Finance Ministers Roger Douglas and Ruth Richardson.

A recession began after the 1987 share market crash and caused unemployment to reach 10% in the early 1990s. Subsequently the economy recovered and New Zealand's unemployment rate reached a record low of 3.4% in the December 2007 quarter, ranking fifth from twenty-seven OECD nations with comparable data. In 2009, New Zealand's economy ranked as the fifth freest in the world according the Heritage Foundation's Index of Economic Freedom. The current government's economic objectives are centred on pursuing free-trade agreements and building a "knowledge economy". On 7 April 2008, New Zealand and China signed the New Zealand China Free Trade Agreement, the first such agreement China has signed with a developed country.

Ongoing economic challenges for New Zealand include a current account deficit of 2.9% of GDP, slow development of non-commodity exports and tepid growth of labour productivity. New Zealand has experienced a series of "brain drains" since the 1970s, as well as educated youth leaving permanently for Australia, Britain or the United States.

"Kiwi lifestyle" and family/whanau factors motivates some of the expatriates to return, while career, culture, and economic factors tend to be predominantly 'push' components, keeping these people overseas. In recent years, however, a brain gain brought in educated professionals from poor countries, as well as Europe, as permanent settlers.

In 2003 New Zealand decriminalised the sex trade, a move which has attracted international attention. Since 2000, New Zealand's fashion industry has grown significantly, doubling exports within a ten year period, according to *The Economist* magazine.

The nation now has "a vibrant and steadily expanding fashion industry, with some 50 established labels, up from a handful ten years ago, half of which sell abroad." Much of this activity is based in Auckland. Clothing exports in 2007 were $315 million, up from $194 million ten years earlier. This is a remarkable turnabout for a nation which has had a reputation for lackluster fashion – "Visiting diplomats have remarked upon the penchant among New Zealand women for short haircuts, backpacks and sensible shoes... One ambassador accused them of dressing like soldiers; another said they looked as though they were going to a funeral."

### *Energy*

In 2008, oil, gas and coal generated approximately sixty-nine percent of New Zealand's gross energy supply and thirty-one per cent was generated from renewable energy, primarily hydroelectric power and geothermal power.

### *Agriculture*

Agriculture has been and continues to be the main export industry in New Zealand. In the year to June 2009, dairy products accounted for 21% ($9.1 billion) of total merchandise exports, and the largest company of the country, Fonterra, a dairy cooperative, controls almost one-third of the international dairy trade. Other agricultural items were meat 13.2%, wood 6.3%, fruit 3.5% and fishing 3.3%.

New Zealand also has a thriving wine industry, which had a bumper year in 2007; wine became New Zealand's "12th most valuable export" in that year, overtaking wool exports. Cows and sheep are rarely housed, but are sometimes fed small quantities of supplements such as hay and silage, particularly in winter. Pigs are usually kept indoors, either in gestation crates, farrowing crates, fattening pens, or group housing. In 1984, the New Zealand Labour Party ended all farm subsidies.

## Demography

### *Ethnicity and Immigration*

New Zealand has a population of about 4.3 million, of which approximately 78% identify with European ethnic groups. The term Pakeha usually refers to New Zealanders of European descent, although some reject this appellation, and some Maori use it to refer to all non-Maori and non-Polynesian New Zealanders. Most European New Zealanders are of British and Irish ancestry, although there has been significant Dutch, Dalmatian, Italian, and German immigration together with indirect European immigration through Australia, North America, South America and South Africa.

According to the 2006 census projections, by 2026 European children will make up 64% of all New Zealand children, compared with 73% in 2006. Maori children will make up 29%, from 24% in 2006, and Asian and Pacific children will make up about 18% each, compared with 9% and 12% in 2006, respectively. The fertility rate as of March 2009 was 2.2 per woman, compared to approximately 2 for the previous 30 years, with the total number of births higher than at any point since 1961. A second fertility estimate was 2.02 children per woman. The fertility rate is expected to decline over the next forty years, according to one estimate.

The life expectancy of a child born in 2008 was 82.4 years for a girl, and 78.4 years for a boy. Life expectancy at birth (males and females) is forecast to increase from 80 years to 85 years in 2050. Further, infant mortality is expected to decline substantially from 2009 to 2050. While the overall population is expected to grow to 5,349,000 in 2050, the median age (half younger, half older) will rise from 36 years in 2009 to 43 years in 2050 and the percentage of people sixty years of age and older will rise from 18% (2009) to 29% (2050).

### *Languages*

Until 1987, English was New Zealand's only official language, and remains predominant in most settings; Maori became an official language under the 1987 Maori Language Act and New Zealand Sign Language under the New Zealand Sign Language Act 2006.

The two official spoken languages are also the most widely used; English is spoken by 98% of the population and Maori by 4.1%. Samoan is the most widely spoken non-official language (2.3%), followed by French, Hindi, Yue and Northern Chinese. New Zealand has an adult literacy rate of 99%, and 14.2% of the adult population has a bachelor's degree or higher. For 30.4% of the population, some form of secondary qualification is their highest, while 22.4% of New Zealanders have no formal qualification.

### *Religion*

According to the 2006 census, Christianity is the predominant religion in New Zealand, held by 55.6% of the population, a decrease from 60.6% at the 2001 census. Another 34.7% indicated that they had no religion, up from 29.6% in 2001, and around 4% affiliated with other religions. The main Christian denominations are Anglicanism, Roman Catholicism, Presbyterianism and Methodism. There are also significant numbers who identify themselves with Pentecostal and Baptist churches and with the LDS (Mormon) church. The New Zealand-based Ratana church has adherents among Maori. According to census figures, other significant minority religions include Hinduism, Buddhism, and Islam.

### *Other Statistics*

In a survey of the OECD amongst 30 democratic nations, New Zealand ranked an above-average 8th place in terms of the happiness of its populace (defined by the averaged responses to questions about personal contentment and positive feelings experienced recently) even though the country was noted as ranking relatively low amongst the surveyed nations in personal wealth (defined by averaged personal income).

New Zealand is a predominantly urban country, with 72% of the population living in 16 main urban areas and 53% living in the four largest cities of Auckland, Christchurch, Wellington, and Hamilton.

## Culture

### *Overview*

While New Zealand is culturally and linguistically part of Polynesia, forming the south-western anchor of the Polynesian Triangle, much of contemporary New Zealand culture is derived from British roots. It also includes significant influences from American, Australian and Maori cultures, along with those of other European cultures and – more recently – non-Maori Polynesian and Asian cultures. Large festivals in celebration of Diwali and Chinese New Year are held in several of the larger centres. The world's largest Polynesian festival, Pasifika, is an annual event in Auckland.

Cultural links between New Zealand and the United Kingdom are maintained by a common language, sustained migration from the United Kingdom, and many young New Zealanders spending time in the United Kingdom on their "overseas experience" (OE). The music and cuisine of New Zealand are similar to that of Australia, Canada, UK, and the US, although both have distinct New Zealand and Pacific qualities.

Maori culture has undergone considerable change since the arrival of Europeans; in particular the introduction of

Christianity in the early 19th century brought about fundamental change in everyday life. Nonetheless the perception that most Maori now live similar lifestyles to their Pakeha neighbours is a superficial one.

In fact, Maori culture has significant differences, for instance the important role which the marae and the extended family continue to play in communal and family life. As in traditional times, Maori habitually perform *karakia* to ensure the favourable outcome of important undertakings, but today the prayers used are generally Christian. Maori still regard their allegiance to tribal groups as a vital part of personal identity, and Maori kinship roles resemble those of other Polynesian peoples.

As part of the resurgence of Maori culture that came to the fore in the late 20th century, the tradition-based arts of *kapa haka* (song and dance), carving and weaving are now more widely practiced, and the architecture of the marae maintains strong links to traditional forms. Maori also value their connections to Polynesia, as attested by the increasing popularity of waka ama (outrigger canoe racing), which is now an international sport involving teams from all over the Pacific.

### *Te reo Maori*

Use of the Maori language (*te reo Maori*) as a living, community language remained only in a few remote areas in the post-war years, but is currently undergoing a process of revitalisation, thanks in part to Maori language immersion schools and two Maori Television channels. These are the only nationwide television channels to have the majority of its prime-time content delivered in Maori, primarily because only 4% of the population speak the language. However, partly in recognition of the importance of Maori culture to New Zealand, the language was declared one of New Zealand's official languages in 1987.

### *Film Industry*

Although films have been made in New Zealand since the 1920s, it was only from the 1970s that New Zealand films began to be produced in significant numbers. Films such as

*Sleeping Dogs* and *Goodbye Pork Pie* achieved local success and launched the careers of actors and directors including Sam Neill, Geoff Murphy and Roger Donaldson. In the early 1990s, New Zealand films such as Jane Campion's Academy Award-winning film *The Piano*, Lee Tamahori's *Once Were Warriors* and Peter Jackson's *Heavenly Creatures* began to garner international acclaim.

In the late 1990s and early 2000s, Jackson filmed *The Lord of the Rings* film trilogy in New Zealand, using mostly New Zealand crew and extras. *Whale Rider*, originally a novel by Witi Ihimaera, was produced in 2002 and received recognition from various festivals and awards. New Zealand features as a primary or additional location for many international productions, examples include *The Lion, the Witch and the Wardrobe*, *Bridge to Terabithia* and *Kaho Naa... Pyaar Hai*.

### *Media*

The New Zealand media industry is dominated by a small number of companies, most of which are foreign-owned, although the state retains ownership of some television and radio stations. New Zealand television broadcasts mostly American and British programming, along with a small number of Australian and New Zealand shows.

The Broadcasting Standards Authority and the New Zealand Press Council can investigate allegations of bias and inaccuracy in the broadcast and print media. Combined with New Zealand's libel laws, this means that the New Zealand news media is fairly tame by international standards, but also reasonably fair and impartia. New Zealand receives high rankings in press freedom. Between 2003 and 2008, Reporters Without Borders has consistently ranked the country in the top twenty, placing it seventh in 2008.

### *Sports*

Sport has a major role in New Zealand's culture, with the unofficial national sport of rugby union being particularly influential. Other popular participatory sports include cricket, bowls, netball, soccer, motorsports, golf, swimming and tennis. New Zealand has strong international teams in several sports including rugby union, netball, cricket, rugby league, and softball. New Zealand also has traditionally done well in the sports of rowing, yachting and cycling. The country is internationally recognised for performing well on a medals-to-population ratio at Olympic Games and Commonwealth Games.

Rugby union, commonly referred to as *rugby*, is closely linked to the country's national identity. The national rugby team, the All Blacks, has the best win to loss record of any national team, and is well known for the haka (a traditional Maori challenge) performed before the start of international matches.

Rugby league is also widely played in New Zealand. The New Zealand Warriors compete in the Australian NRL competition, and in 2008 the national side, the Kiwis, won the Rugby League World Cup. Cricket was introduced to New Zealand in the 1800s and is reputedly the second most popular sport in the country, with one source stating there are 98,000 registered cricket players. The New Zealand team is known as the Blackcaps and the national women's team is the White Ferns. Horse racing is a popular spectator sport which has spawned such national icons as Cardigan Bay and Phar Lap, and was part of the traditional "Rugby, Racing and Beer" culture.

New Zealand is also well known for its extreme sports and adventure tourism. Its reputation in extreme sports extends from the establishment of the world's first commercial bungy jumping site at Queenstown in the South Island in November 1988. The country also has a strong mountaineering tradition, with the country's most famous climber being the late Sir Edmund Hillary, jointly with Tenzing Norgay the first person to reach the summit of Mount Everest.

## Geographical Zone

The idea of a Geographical zone was first hypothesized by the ancient Greek scholar Aristotle in an attempt to classify the world's climatic zones. He determined three types of climatic zones based upon location in relation to the equator. While his mapping was oversimplified due to limited geographical knowledge available at the time, his general understanding was correct. As knowledge of the earth's geography improved, the geographical zones increased to five, based upon each of five main latitude regions divided by the major circles of latitude.

Today, there are several climate classifications systems used, the four most common being: The Koppen climate classification; Trewartha climate classification scheme (based upon the Koppen); the Aridity index; and the Holdridge life zones system. However, climate classification is a multidimensional concept, considering many environmental variables. While no classification system is absolute, they play a key role in discerning the geographic distribution and interactions of existing climatic elements. Individually and together they provide important data for climate awareness. They serve as educational tools for understanding climate in the world's most distant regions.

## Circles of Latitude

The equator is the imaginary line on the Earth's surface that is equidistant from both the North Pole and South Poles and divides the Earth into the Northern Hemisphere and the Southern Hemispheres. It is from this point that latitude is calculated, in other words, it is the 0° latitude line. Including the equator, there are five major circles of latitude, from north to south:

- The Arctic Circle (66° 332 383 N)
- The Tropic of Cancer (23° 262 223 N)
- The Equator (0° latitude)
- The Tropic of Capricorn (23° 262 223 S)
- The Antarctic Circle (66° 332 383 S).

These circles of latitude (excluding the equator) mark the divisions between the five principal geographical zones. Arcs of circles of latitude are sometimes used as boundaries between countries or regions where distinctive natural borders are lacking (such as in deserts), or when an artificial border is drawn as a "line on a map," as happened in Korea.

### *Zone Descriptions*

Each of the five main latitude regions of the earth's surface is said to be a geographical zone, divided by the major circles of latitude. The differences between them relate to climate, and the behaviour of the sun. They are as follows:

- The North Frigid Zone, north of the Arctic Circle
- The North Temperate Zone, between the Arctic Circle and the Tropic of Cancer
- The Torrid Zone, between the Tropics of Cancer and Capricorn
- The South Temperate Zone, between the Tropic of Capricorn and the Antarctic Circle
- The South Frigid Zone, south of the Antarctic Circle.

In the Torrid Zone, also known as the Tropics, the sun is directly overhead at least once during the year—at the edges of the tropics this occurs at the summer solstice, and over the equator, at the equinoxes.

This is the hottest part of the earth, and there are two annual seasons: a dry and a wet. The Torrid Zone includes most of Africa, southern India, southern Asia, Indonesia, New Guinea, northern Australia, southern Mexico, Central America, and northern South America.

In the two Temperate Zones, the sun is never directly overhead, and the climate is mild, generally ranging from warm to cool. The four annual seasons, Spring, Summer, Autumn and Winter occur in these areas. The North Temperate Zone includes Great Britain, Europe, northern Asia, North America, and northern Mexico. The South Temperate Zone includes southern Australia, New Zealand, southern South America, and South Africa. The edges of the Temperate Zones are also sometimes referred to as the Subtropics, Subarctic, and Subantarctic.

The two Frigid Zones, or polar regions, experience the midnight sun and the polar night for part of the year—the edge of the zone experiences one day at the solstice when the sun doesn't rise or set for 24 hours, while in the centre of the zone (the pole), the day is literally one year long, with six months of daylight and six months of night. The Frigid Zones are the coldest parts of the earth, and are covered with ice and snow. The North Frigid Zone (the Arctic) includes northern Canada and Alaska, Greenland, northern Scandinavia, northern Russia, and the Arctic ice. The South Frigid Zone (The Antarctic) is filled by the continent of Antarctica; the next closest mainland is the southern tip of Chile and Argentina, followed by New Zealand.

While both the North and South Poles are in frigid zones, there is a vast difference in temperature between the two. The North Pole lies at sea level in the middle of the Arctic Ocean, which acts as a reservoir of heat. The sea ice at the North Pole is typically around two or three meters thick, though there is considerable variation and occasionally the movement of floes exposes clear water.

The South Pole sits atop a featureless, windswept, icy plateau at an altitude of 9,306 feet (2,836 m), about 8,000 miles (13,000 km) from the nearest sea at McMurdo Sound. The ice is estimated to be about 9,000 feet (2,700 m) thick at the Pole. Winter temperatures at the North Pole can range from about “45ºF (“43°C) to “15°F (“26°C), perhaps averaging around “30ºF (“34°C). Summer temperatures (June, July and August) average around the freezing point (0°C, 32°F). In comparison, the winter temperature at the South Pole remains steady at around “85°F (“65°C). The highest temperature ever recorded at the Amundsen-Scott South Pole Station is 7.5°F (“13.6°C), and the lowest is “117.0°F (“82.8°C).

## *History*

The idea of a geographical zone was first hypothesized by the ancient Greek scholar Aristotle. He said that the earth was divided into three types of climatic zones, based on their distance from the equator.

Thinking that the area near the equator was too hot for habitation, Aristotle dubbed the region around the equator (from 23.5° N to 23.5° S) as the “Torrid Zone.” He reasoned that from the Arctic Circle to the pole was permanently frozen. He called this uninhabitable zone the “Frigid Zone.” The only area that Aristotle believed was livable was the “Temperate Zone,” lying between the “Frigid Zone” and the “Torrid Zone.” One of the reasons Aristotle believed that the Temperate Zone was the best for life possibly stems from the fact that he lived in that zone.

As knowledge of the earth's geography improved, a second "Temperate Zone" was discovered south of the equator, and a second "Frigid Zone" was discovered around the Antarctic. It is understood today that Aristotle's map was vastly oversimplified, although the general idea was correct.

## *Modern Climatic Systems*

Climate classification systems are ways of classifying the world's climates. A classification may correlate closely with a biome category, as climate is a major influence on biological life in a region. The most commonly used climate map is the Koppen climate classification scheme. Others in use are the Aridity index and the Holdridge Life Zone Classification System.

The Koppen climate classification, developed by German climatologist and amateur botanist Wladimir Koppen (1846-1940) is based on the concept that native vegetation is the best expression of climate; thus, climate zone boundaries were selected with vegetation distribution in mind.

The system combines average annual and monthly temperatures and precipitation, and the seasonality of precipitation. The Koppen climate classification scheme divides the climates into five main groups and several types and subtypes. The Koppen classification includes climate regimes such as rainforest, monsoon, tropical savanna, humid subtropical, humid continental, oceanic climate, Mediterranean climate, continental steppe, subarctic climate, tundra, polar ice cap, and desert.

The Koppen system, however, has its drawbacks. A modified version of the Koppen system known as the Trewartha climate classification scheme attempts to redefine the broad climatic groups in such a way as to be closer to vegetational zoning, especially in the United States. Under the standard Koppen system western Washington and Oregon are classed into the same climate as southern California, even though the two regions have strikingly different vegetation. It also classes southern New England into the same climate as the Gulf Coast. Trewartha's modifications sought to reclassify the Pacific Northwest seaboard as a different climate from California, and New England from the Gulf Coast.

The Holdridge life zones system is a global bioclimatic scheme for the classification of land areas. It was first published by Leslie Holdridge in 1947, and updated in 1967. It is a relatively simple system based on few empirical data, giving objective mapping criteria. A basic assumption of the system is that both soil and climax vegetation can be mapped once climate is known.

While it was first designed for tropical and subtropical area, the system applies globally. The system has been shown to fit tropical vegetation zones, Mediterranean zones, and boreal zones, but is less applicable to cold oceanic or cold arid climates where moisture becomes the determining factor. The system has found a major use in assessing the possible changes in natural vegetation patterns due to global warming.

# Bibliography


Belden C.: *Landscapes of the Sacred: Geography and Narrative in American Spirituality*, New York: Paulist Press, 1988.

Berry, J.K. : *Beyond Mapping, Concepts, Algorithms and Issues in GIS*, Fort Collins, CO: GIS World Books, 1993.

Bolstad, P. : *GIS Fundamentals, A first text on Geographic Information Systems, Second Edition*, White Bear Lake, MN, Eider Press, 2005.

Cronon, William: *Changes in the Land: Indians, Colonists, and the Ecology of New England.* New York: Hill and Wang, 1983.

Davis, Mike: *Ecology of Fear: Los Angeles and the Imagination of Disaster*. New York: Vintage, 1999.

Elangovan K. : *GIS, Fundamentals, Applications and Population Geography*, New India Publishing Agency, New Delhi 2006.

Harvey, David: *Explanation in Geography.* London: Edward Arnold, 1969.

Harvey, Francis : *A Primer of GIS, Fundamental Geographic and Cartographic Concepts,* The Guilford Press, 2008.

Jackson, Peter: *Maps of Meaning: An Introduction to Cultural Geography.* London and Boston: Unwin Hyman, 1989.

James S. Duncan: *Writing Worlds: Discourse, Text, and Metaphor in the Representation of Landscape.* London and New York: Routledge, 1992.

James, T.: *The Best Poor Man's Country: A Geographical Study of Early Southeastern Pennsylvania*. Baltimore: Johns Hopkins Press, 1972.

King, Anthony D.: *Buildings and Society: Essays on the Social Development of the Built Enviroment*. London: Routledge & Kegan Paul, 1980.

Lewis, Peirce. *New Orleans: The Making of an Urban Landscape.* Cambridge, MA: Ballinger Publishing Co., 1976. Annotation

Liam Kennedy: *Urban Space and Representation.* Sterling, VA: Pluto Press, 2000.

Lung, Y., and van Tulder, R. : *Cars, Carriers of Regionalism*, Palgrave Macmillan, New York, 2004.

McDonnell, R.A. : *Principles of Geographical Information Systems*, Oxford University Press, Oxford, 1998.

Michael P.: *The Making of the American Landscape.* Boston: Unwin Hyman, 1990.

Michael, H.: *City Form and Natural Process: Towards a New Urban Vernacular.* London: Croom Helm, 1984.

Peter, G.: *The Slow Plague: A Geography of the AIDS Pandemic.* Oxford and Cambridge, MA: Blackwell Publishers, 1993.

Ray Allen: *The Far Western Frontier, 1830-1860.* The New American Nation Series. New York: Harper, 1956.

Richard SCASE : *Managing Creativity,* Open University Press, 2000.

Stephen, D.: *Fields of Vision: Landscape Imagery and National Identity in England and the United States.* Cambridge: Polity, 1993.

Tulder, R. : *Cars, Carriers of Regionalism*, Palgrave Macmillan, New York, 2004.

Victor, A. : *Geography of Art and Culture*, Contributions to Economic Analysis, Elsevier, 2004.

Watkins, Eric: *The Middle Eastern Environment*, London, The British Society for Middle Eastern Studie, 1995.

William G. BOWEN : *Performing Arts, The Economic Dilemma,* Thousand Oaks, SAGE, London, 2002.

Worboys, Michael, and Matt Duckham. : *GIS, a Computing Perspective*, Boca Raton, CRC Press, 2004.

# Index

## E



## F



## G



## H



## I



## L

## R



## S



## T



## U



## V



## W



□□□